Novel Insights and Advances in Steels and Cast Irons

Novel Insights and Advances in Steels and Cast Irons

Guest Editors

Annalisa Fortini
Chiara Soffritti

MDPI

Basel • Beijing • Wuhan • Barcelona • Belgrade • Novi Sad • Cluj • Manchester

Guest Editors
Annalisa Fortini
Department of Engineering
University of Ferrara
Ferrara
Italy

Chiara Soffritti
Department of Engineering
University of Ferrara
Ferrara
Italy

Editorial Office
MDPI AG
Grosspeteranlage 5
4052 Basel, Switzerland

This is a reprint of the Special Issue, published open access by the journal *Metals* (ISSN 2075-4701), freely accessible at: https://www.mdpi.com/journal/metals/special_issues/571SWR3I81.

For citation purposes, cite each article independently as indicated on the article page online and as indicated below:

Lastname, A.A.; Lastname, B.B. Article Title. *Journal Name* **Year**, *Volume Number*, Page Range.

ISBN 978-3-7258-7496-5 (Hbk)
ISBN 978-3-7258-7497-2 (PDF)
https://doi.org/10.3390/books978-3-7258-7497-2

Cover image courtesy of Annalisa Fortini

Contents

metals MDPI

Editorial

Novel Insights and Advances in Steels and Cast Irons

Annalisa Fortini * and Chiara Soffritti

Department of Engineering, University of Ferrara, Via Saragat 1/E, 44122 Ferrara, Italy; chiara.soffritti@unife.it
* Correspondence: annalisa.fortini@unife.it; Tel.: +39-0532-974914

1. Introduction and Scope

The manufacturing sector relies extensively on ferrous alloys, with steels and cast irons serving as essential materials for structural and functional components across a wide range of industries, including automotive, energy generation, and civil engineering.

Cast irons continue to play a significant industrial role due to their excellent castability, cost-effectiveness, and intrinsic properties, such as vibration damping and wear resistance, provided by their graphite microstructure. These features make them ideal for complex, near-net-shape components. However, the rapid technological evolution, especially in fields such as Concentrated Solar Power (CSP) plants, high-speed motors, and advanced automotive bodywork, places ever-increasing demands on material performance. This ongoing progress calls for the development of innovative alloys with improved high-temperature corrosion resistance, enhanced processing efficiency, and mechanical stability under severe deformation.

Alongside cast irons, steels continue to represent the cornerstone of numerous industrial and structural applications. Continuous research efforts are crucial for developing high-performance materials in both families, aimed at improving in-service performance under increasingly stringent conditions. In particular, steels constituting the core focus of this Special Issue are subject to intense research directed towards (i) optimizing mechanical properties through controlled microstructural evolution, (ii) mitigating degradation mechanisms in harsh environments, including hot corrosion by molten salts at elevated temperatures and addressing complex paint failures like filiform corrosion on coated substrates and (iii) advancing manufacturing precision and structural integrity, through techniques including additive manufacturing or novel welding methods designed to minimize structural defects.

Within this framework, the present Special Issue collects ten high-quality research papers, addressing these areas of investigation, providing valuable theoretical and experimental insights that advance the understanding of ferrous alloy behavior and promote sustainable progress in material performance and manufacturing technologies.

2. Contributions

The contributions encompass a wide range of topics, including material behavior under severe plastic deformation, optimization of inclusions in free-cutting steels, microstructural stability in heat-resistant alloys, and innovative approaches to corrosion prevention.

The first group of papers explores the interaction between ferrous alloys and their service environments, focusing on corrosion and oxidation under aggressive or functional conditions. The studies address key degradation mechanisms—such as hot corrosion in molten salts and filiform corrosion in coated steels—and propose mitigation strategies including eco-friendly inhibitors and combined thermomechanical-surface treatments.

 https://doi.org/10.3390/met15111240

Abu-warda et al. (Contribution 1) investigated the corrosion behavior of laser powder bed fusion (L-PBF) 316L stainless steel exposed to various molten salts intended for thermal energy storage in concentrating solar power (CSP) plants. Corrosion tests conducted at 650 °C and 700 °C revealed that salt composition had a greater influence than temperature. Chloride-based salts produced the most severe degradation, followed by carbonate- and nitrate/nitrite-based mixtures. Surface and cross-sectional analyses by X-ray diffractometry (XRD) and scanning electron microscopy (SEM) equipped with energy dispersive spectroscopy (EDS) identified $LiFeO_2$ and $LiCrO_2$ as key corrosion products in carbonate media, while chloride salts induced active oxidation through a chlorine cycle. The results highlight that L-PBF 316L is unsuitable for CSP systems using chloride salts above 650 °C without protective measures.

Cristoforetti et al. (Contribution 2) examined the effect of disodium sebacate (SB), an environmentally friendly corrosion inhibitor, in mitigating filiform corrosion on acrylic-coated carbon steel substrates. Electrochemical impedance spectroscopy, polarization tests, and a dedicated filiform corrosion-simulating electrochemical setup revealed that SB acts primarily as an anodic inhibitor in near-neutral conditions, achieving an inhibition efficiency of about 98%. When incorporated into the coating primer (1 wt. %), SB reduced corrosion-creep propagation by about 35% during accelerated weathering. The molecule mitigated cathodic activity under alkaline, aerated environments typical of filiform corrosion and formed a hydrophobic interfacial layer that slows water ingress. These findings confirm sebacate's promise as an environmentally benign additive for enhancing the durability of organic-coated steels against localized delamination phenomena.

Vontorová et al. (Contribution 3) explored the enhancement of the mechanical properties of low-carbon DC03 steel through severe plastic deformation via the dual-rolling equal-channel extrusion (DRECE) method, followed by hot-dip zinc galvanizing. The effects on microstructure, hardness, and corrosion behavior were evaluated. The DRECE process enhanced hardness by up to 20%, without significantly affecting the uniformity or thickness of the zinc coating (of about 70–85 μm), while microstructural analyses confirmed the presence of a typical Fe–Zn intermetallic layer. Electrochemical testing, based on Tafel extrapolation, revealed that DRECE treatment had a negligible impact on corrosion potential and current, while galvanizing drastically improved corrosion resistance. Hence, the combined DRECE and galvanizing route provides mechanically strengthened, corrosion-protected steels suitable for structural and automotive applications.

Another line of investigation explores how thermal treatments and phase transformation control can be leveraged to optimize the performance of steels. By linking heat treatment parameters to wear resistance, hardness, and microstructural evolution, the works collectively demonstrate the value of precise phase control in improving tribological performance, mechanical balance, and thermal stability.

Li et al. (Contribution 4) investigated the hot deformation behavior and optimized processing parameters of a Cu-alloyed $4Cr_{16}MoCu$ martensitic stainless steel through Gleeble-3500 compression tests conducted between 900 °C and 1150 °C and at strain rates of 0.001–1 s^{-1}. The addition of 1 wt. % Cu enhanced precipitation hardening and corrosion resistance but significantly affected hot workability. Flow-stress data were modeled using a strain-compensated Arrhenius constitutive equation, yielding an activation energy of $\approx$470 kJ mol^{-1} and excellent agreement with experiments. Based on microstructural analysis and the hot processing map, the authors determined the optimal forging window at 1125–980 °C and 0.1 s^{-1}, which was able to promote dynamic recrystallization and fine grains, while strain rates > 0.18 s^{-1} caused flow instability. The results provide quantitative guidance for forging Cu-bearing martensitic stainless steels.

Wang et al. (Contribution 5) studied the influence of B2 phase precipitation on the hot ductility of $Fe_{22}Mn_9Al_{0.6}C$ low-density steel, a candidate material for lightweight automotive components. High-temperature tensile tests conducted between 800 °C and 950 °C revealed a marked drop in ductility and strength within 850–900 °C, associated with the dissolution of κ-carbide and the emergence of the ordered B2 phase, which occurred as temperature increased. The B2 phase is believed to cause crack generation during the hot-forging process. XRD, SEM, and transmission electron microscope (TEM) analyses confirmed the sequential phase transformations, revealing that the precipitate promotes intergranular cracking and embrittlement during hot deformation. The findings demonstrate that avoiding the B2 precipitation temperature range is critical to maintain hot workability and prevent cracking in FeMnAlC low-density steels.

In the study of Qiao et al. (Contribution 6), the effects of solution treatment temperature on the microstructure, mechanical properties, and wear resistance of Fe–25Mn–0.37C–3.69Cr high-manganese steel were examined. Samples treated between 900 °C and 1050 °C exhibited grain coarsening after initial refinement, with optimal recrystallization and uniform austenitic grains at 950 °C. The steel solution-treated at 950 °C achieved the best balance of tensile strength (574 MPa), yield strength (268 MPa), elongation (88%), and wear resistance, associated with enhanced work-hardening during friction. The friction coefficient reached a minimum of 0.273 at this temperature due to oxidation-assisted wear. The dominant wear mechanism transitioned from abrasive and fatigue wear to adhesive wear with increasing solution temperature. These findings provide insights for optimizing heat treatment of alloyed high-manganese steels.

Finally, Timotijević et al. (Contribution 7) provided an insight into the short-term microstructural evolution of HP40Nb austenitic heat-resistant steel, a centrifugally cast alloy used in reformer furnaces, exposed to 950 °C, 1050 °C, and 1150 °C for 2 h and 8 h, simulating overheating conditions in petrochemical reformer tubes. Optical, SEM/EDS, and XRD analyses revealed an as-cast austenitic dendritic matrix containing primary eutectic-like $M_{23}C_6$ and MC carbides. Increasing temperature and holding time promoted carbide coarsening, spheroidization of primary carbides, and progressive dissolution of secondary $Cr_{23}C_6$ and NbC precipitates into the matrix. The highest hardness occurred after 8 h at 1150 °C, attributed to partial carbide dissolution and carbon diffusion within austenite. However, exposure at 1150 °C for 8 h resulted in a reduction in secondary carbides due to dissolution, leading to the formation of carbide-free zones. The findings clarify carbide transformation mechanisms under transient overheating and underscore the importance of microstructural monitoring for extending the service life of reformer tubes in extreme environments.

The final group of articles delves into the broader processing-structure-property relationships that govern the mechanical and functional performance of steels in industrial contexts. Topics include microalloying strategies (e.g., Cu, Bi, Te), solidification behavior, inclusion modification, and hot working behavior. Studies in this section explore inverse segregation, dynamic recrystallisation, B2 intermetallic suppression, and machinability enhancement through inclusion engineering. Special emphasis is placed on translating laboratory insights to real-world applications, highlighting how alloy design, casting control, and thermomechanical routes interact to define the final performance of advanced steel components.

The study by Liu et al. (Contribution 8) proposed a novel indentation-free resistance spot welding method for joining SUS301L stainless steel sheets in paint-free rail vehicle applications, utilizing a 1.5 mm steel ball as an intermediate filler. The latter was intended to redistribute current density and confine heat generation within the nugget, thereby minimizing surface indentation. Systematic variation of welding current (7.0–8.5 kA), time (120–210 ms), and pressure (0.35 MPa) revealed optimal joint strength (6414 N) at 8.0 kA and 150 ms, without surface ablation or spatter. The nugget exhibited columnar austenite and a narrow heat-affected zone, similar to conventional RSW. Under optimized conditions, indentation depth was <1% of plate thickness, confirming that the proposed method enables mechanically robust, visually defect-free welds suitable for lightweight stainless-steel body structures.

Wang et al. (Contribution 9) examined the effects of environmentally friendly alloying elements bismuth (Bi) and tellurium (Te) on the inclusion characteristics of sulfur-containing free-cutting steels. Three compositions, S-only, S and Bi, and S, Bi and T, were prepared and analyzed through metallography, SEM/EDS, and 3D electrolytic etching. The results revealed that Bi and Te significantly modify MnS inclusions, transforming elongated class II morphologies into fine flake- or spindle-shaped class III inclusions with lower aspect ratios (<4), thereby improving machinability and reducing anisotropy. The Te addition promoted MnS-MnTe composite inclusions of diverse morphologies (heart-, droplet-, or butterfly-shaped). Both elements decreased the overall inclusion count and homogenized their distribution. The combined Bi-Te modification offers a sustainable route to lead-free free-cutting steels with enhanced cutting performance and isotropy.

Finally, Gao et al. (Contribution 10) investigated the microstructural evolution, texture development, and strengthening mechanisms of Fe–3.3 wt. % Si non-oriented electrical steel subjected to double cold rolling and final annealing. Microstructure and texture were characterized using optical microscopy (OM,) XRD, and quasi-in situ electron backscattered diffraction (EBSD) techniques to reveal recrystallization behavior. The double cold-rolling route effectively suppressed the detrimental γ-fiber and promoted η-fiber texture formation, optimizing the texture factor and magnetic response. At reduction ratios of 50–65%, the steel exhibited enhanced magnetic induction (B50 = 1.698 T), reduced iron loss (P10/400 = 21.84 W/kg), and increased yield strength (Rp0.2 = 578 MPa) compared with single-stage rolling. The combined dislocation-strengthening and texture-control strategy offers a pathway for high-strength, low-loss silicon steels suitable for next-generation electrical applications.

3. Conclusions and Outlook

The research compiled in this Special Issue effectively demonstrates the dynamic and multifaceted nature of current advancements in ferrous metallurgy. The collected contributions underscore the continued industrial and scientific relevance of advanced steels and cast irons, particularly in terms of their mechanical, tribological, and corrosion-related performance.

The findings demonstrate that significant property enhancements can be achieved through careful control of processing conditions and in-depth analyses of microstructural evolution. Steels and cast irons maintain a high degree of industrial and scientific relevance, with potential for enhanced performance. Altogether, the contributions gathered in this Special Issue offer valuable insight and theoretical support for the design of robust, cost-effective, and high-performance components, serving as a reference for both academic research and industrial innovation.

Conflicts of Interest: The authors declare no conflicts of interest.

List of Contributions:

1. Abu-warda, N.; García-Rodríguez, S.; Torres, B.; Utrilla, M.V.; Rams, J. Effect of Molten Salts Composition on the Corrosion Behavior of Additively Manufactured 316L Stainless Steel for Concentrating Solar Power. *Metals* **2024**, *14*, 639. https://doi.org/10.3390/met14060639.

2. Cristoforetti, A.; Rossi, S.; Deflorian, F.; Fedel, M. Unlocking the Potential of Sebacate: Investigating Its Role in the Inhibition of Filiform Corrosion on Organic Coated Steel. *Metals* **2024**, *14*, 623. https://doi.org/10.3390/met14060623.
3. Vontorová, J.; Novák, V.; Váňová, P. Low-Carbon Steel Formed by DRECE Method with Hot-Dip Zinc Galvanizing and Potentiodynamic Polarization Tests to Study Its Corrosion Behavior. *Metals* **2024**, *14*, 993. https://doi.org/10.3390/met14090993.
4. Li, J.; Li, L.; Wu, Z.; Zeng, T.; Wu, X. Hot Deformation Behavior and Optimization of Processing Parameters for 4Cr16MoCu Martensitic Stainless Steel. *Metals* **2025**, *15*, 373. https://doi.org/10.3390/met15040373.
5. Wang, J.; Man, T.; Zhou, Y.; Wei, X.; Dong, H. Effect of B2 Precipitation on Hot Ductility of Fe–22Mn–9Al–0.6C Low-Density Steel. *Metals* **2024**, *14*, 724. https://doi.org/10.3390/met14060724.
6. Qiao, X.; Li, B.; Han, X.; Zhang, X.; Yang, X. Effect of Solution Treatment on Mechanical Properties and Wear Resistance of Alloyed High-Manganese Steel. *Metals* **2025**, *15*, 937. https://doi.org/10.3390/met15090937.
7. Timotijević, M.; Erić Cekić, O.; Janjatović, P.; Kremenović, A.; Rosić, M.; Stopic, S.; Rajnović, D. Insight to the Microstructure Analysis of a HP Austenitic Heat-Resistant Steel Under Short-Term High-Temperature Exposure. *Metals* **2025**, *15*, 486. https://doi.org/10.3390/met15050486.
8. Liu, Y.; Xie, Y.; Sun, X.; Sun, L.; Wang, N.; Ma, X.; Meng, X.; Huang, Y. Indentation-Free Resistance Spot Welding of SUS301L Stainless Steel. *Metals* **2024**, *14*, 1178. https://doi.org/10.3390/met14101178.
9. Wang, X.; Zhang, H.; Wang, J.; Zhu, R.; Zhu, Y.; Lu, F.; Li, J.; Jiang, Z. Effect of Bismuth and Telluride on the Inclusions of Sulfur Free-Cutting Steel. *Metals* **2023**, *13*, 486. https://doi.org/10.3390/met13030486.
10. Gao, Y.; Xu, Y.; Chen, H.; Yuan, B.; Gao, Z.; Zhou, L. Dislocation Strengthening and Texture Evolution of Non-Oriented Fe-3.3 Wt% Si Steel in Double Cold Rolling. *Metals* **2024**, *14*, 1127. https://doi.org/10.3390/met14101127.

metals

MDPI

Article

Effect of Molten Salts Composition on the Corrosion Behavior of Additively Manufactured 316L Stainless Steel for Concentrating Solar Power

Najib Abu-warda *, Sonia García-Rodríguez, Belén Torres, María Victoria Utrilla and Joaquín Rams

Departamento de Matemática Aplicada, Ciencia e Ingeniería de Materiales y Tecnología Electrónica, Universidad Rey Juan Carlos, Tulipán s/n, 28933 Móstoles, Spain; sonia.garcia@urjc.es (S.G.-R.); belen.torres@urjc.es (B.T.); victoria.utrilla@urjc.es (M.V.U.); joaquin.rams@urjc.es (J.R.)

* Correspondence: najib.abuwarda@urjc.es; Tel.: +34-914887160

Abstract: The effects of different molten salts on the corrosion resistance of laser powder bed fusion (L-PBF) 316L stainless steel was evaluated at 650 and 700 °C. The samples were characterized via XRD and SEM/EDX after high-temperature corrosion tests to evaluate the corrosion damage to the L-PBF 316L stainless steel caused by the molten salts. The presence of the salts accelerated the corrosion process, the chloride-based salts being the most aggressive ones, followed by the carbonate-based and the nitrate/nitrite-based salts, respectively. The L-PBF 316L did not react strongly with the nitrate/nitrite-based salts, but some corrosion products not found in the samples tested in the absence of salts, such as $NaFeO_2$, were formed. $LiFeO_2$ and $LiCrO_2$ were identified as the main corrosion products in the samples exposed to the carbonate-based molten salts, due to the high activity of Li ions. Their growth produced the depletion of Fe and Cr elements and the formation of vacancies that acted as diffusion paths on the surface of the steel. In the samples exposed to chloride-based molten salts, the attacked area was much deeper, and the corrosion process followed an active oxidation mechanism in which a chlorine cycle is assumed to have been involved.

Keywords: 316L stainless steel; L-PBF; molten salts; corrosion; oxidation; thermal energy storage

1. Introduction

Concentrated solar power (CSP) plants have increased in number in recent decades for decarbonizing the electricity grid due to their ability to integrate cost-effective thermal energy storage (TES) [1–3]. The use of TES materials that accumulate heat during sunshine periods and release it during periods of no solar irradiation has been extensive to overcome the problem of intermittency [4]. In fact, it is expected that 77% of the new CSP plants will integrate TES systems [5]. The next generation of CSP plants is expected to be more efficient in terms of energy production and this is directly related to the maximum working temperature of CSP plants and their capacity to store this energy for a time [6,7].

There are different methods employed for TES, such as latent heat in phase change materials, thermochemical, or solid sensible heat storage [8]. Sensible heat storage based on molten salts is the most used. The salt energy content increases with temperature and, when the material cools, the stored energy is released without any phase change during charge or discharge [9]. In this context, the current CSP plants use a non-eutectic salt mixture of 60 wt.% $NaNO_3$ and 40 wt.% KNO_3 [10]. This commercial salt, commonly known as solar salt, has high heat capacity and low price; however, it also has low thermal stability limits. Its use supposes that the maximum operating temperature is near 550 °C [11]. In this context, other salt mixtures have been considered to overcome this maximum operation limit. Particularly, carbonate-based $32K_2CO_3$-$35Li_2CO_3$-$33Na_2CO_3$ salt (in wt.%) has a higher thermal stability temperature, but the price of Li_2CO_3 limits its use [12, 13]. Other TES materials based on chloride salts, such as 23.9KCl-7.5NaCl-68.6$ZnCl_2$

or 17.8KCl-14NaCl-68.2$ZnCl_2$ have been proposed because they meet the requirements of high thermal stability limits and low price [14]. In addition, chloride salts, such as NaCl and KCl, are available in large amounts, and when mixed with $ZnCl_2$ or $MgCl_2$, eutectic mixtures with low melting points and high stability limits are achieved. However, despite these advantages, chloride-based TES materials have not yet been implemented due to metallurgical problems related to their high corrosiveness [15,16]. For this reason, more efforts are needed to understand the effect of these new TES materials on the high-temperature corrosion behavior of metal alloys used as structural components in CSP plants.

In this context, austenitic stainless steels, and specifically, 316L stainless steel, are the most used structural materials in many industries, mainly due to their good corrosion resistance, which is attributed to the addition of Cr and Mo [17–19], and their good mechanical properties at high temperatures [20]. Moreover, its good oxidation resistance makes 316L stainless steel an excellent candidate to be used in CSP plants and high-temperature parts such as pipelines or hot tanks manufactured with stainless steel [21–23].

In addition, the manufacturing process also plays an important role in the corrosion behavior of 316L ss. In a previous study, the influence of the manufacturing process on corrosion resistance was evaluated in the presence of 17.8KCl-14NaCl-68.2$ZnCl_2$ molten salt [24]. The 316L manufactured by laser powder bed fusion (L-PBF) provided more corrosion resistance than the wrought one, due to the presence of more pathways for Cr diffusion associated with a higher number of defects in the microstructure of the L-PBF samples, facilitating the formation of the protective Cr_2O_3 scale. In this context, additive manufacturing (AM) techniques have attracted great interest in recent years to produce components for different industry sectors.

AM encompasses a set of manufacturing techniques based on the production of components through adding material layer by layer from a three-dimensional (3D) model [25]. In the field of metals, L-PBF, also known as selective laser melting (SLM), is one of the most used AM techniques [26–28]. Among other advantages, these techniques stand out due to the possibility of manufacturing components with more complex geometries than those obtained with casting processes [29].

The design flexibility of AM allows the manufacturing of pieces as single components, avoiding the need to weld different parts [30]. These new AM techniques could be implemented in valves, complex pipes, pumps, or complex heat exchangers in the energy sector [31]. However, the corrosion behavior of alloys manufactured through L-PBF in aggressive environments has not been studied. In this context, this work presents a comparative analysis of the corrosiveness of different TES molten salts, analyzing their effect on L-PBF 316L stainless steel. The results of this research provide helpful information about both the use of these molten salts as TES materials in the next generation of CSP plants and their effect on 316L stainless steel manufactured through L-PBF.

2. Materials and Methods

2.1. Base Material

The austenitic 316L ss produced via L-PBF with a size of 15 × 15 × 3 mm^3 was manufactured in a vertical orientation and was provided by Idonial (Madrid, Spain). The additive manufacturing system was a fiber laser M280 400W (EOS, Krailling, Germany). The samples were manufactured with 1070 nm laser beam wavelength, 0.07 mm laser spot diameter, and 195 W power. In addition, 1083 mm/s scanning speed, 20 µm layer height, and 90 µm hatching distance were used. The composition of the powders used in the L-PBF process was 17.0 Cr, 13.1 Ni, 2.7 Mo, 1.3 Mn, 1.2 Si, ≤0.1 O, ≤0.03 C, and Fe (rest) (in wt.%), and the particle size of the powder was in the range of 20–50 µm.

2.2. Corrosion Tests

Hot corrosion tests were performed to evaluate the effect of salt composition on the 316L stainless steel without any surface treatment. In this context, the as-built samples had an average surface roughness (R_a) of 2.8 ± 0.6 μm. Different salt mixtures were used for this purpose, and their composition and nomenclature are shown in Table 1. The first batch of samples was tested without any salt mixture and was used for comparative purposes. The second batch of samples was exposed to a nitrate/nitrite-based salt mixture, which was selected because it is the most implemented molten salt of the four analyzed and was employed as a reference salt in this work [32]. The $KNaNO_x$ combinations are unstable at the temperature used in this study, but they have been included as a common reference and to study conditions in which there is a punctual overheating of the system. The third batch was exposed to a carbonate-based salt mixture [33], and the last two batches of samples were exposed to two different chloride-based salt mixtures, one rich in $ZnCl_2$ [34] and the other one rich in $MgCl_2$ [35]. The latter three molten salts were investigated because they are promising TES materials for use in CSP plants.

Table 1. Molten salts properties, composition (in wt.%), and nomenclature.

Molten Salt	No Salts	Nitrates/Nitrite	Carbonates	Zn-Chlorides	Mg-Chlorides
Nomenclature	**Oxidation**	**$KNaNO_x$**	**$LiNaKCO_3$**	**NaKZnCl**	**NaKMgCl**
Melting point (°C)	-	142	397	204	380
Stability limit (°C)	-	535	>650	850	>800
KNO_3	-	53.0	-	-	-
$NaNO_3$	-	7.0	-	-	-
$NaNO_2$	-	40.0	-	-	-
K_2CO_3	-	-	32.0	-	-
Li_2CO_3	-	-	35.0	-	-
Na_2CO_3	-	-	33.0	-	-
KCl	-	-	-	23.9	17.8
NaCl	-	-	-	7.5	14.0
$ZnCl_2$	-	-	-	68.6	-
$MgCl_2$	-	-	-	-	68.2

High-temperature corrosion tests were performed in a dry air atmosphere to simulate a punctual failure in the inertization system of a plant using LT 5/12/P330 furnace (Nabertherm GmbH, Lilienthal, Germany), following the standard ISO/FDIS 17224:2014 for high-temperature corrosion tests [36]. Figure 1 shows a scheme of the testing method employed for the corrosion tests; two samples were used for each studied condition. A short exposure time (48 h) was selected, also to simulate the punctual failure in the inertization system. In addition, two temperatures (650 and 700 °C) were employed. These temperatures were selected because the working temperature is expected to rise in the next generation of CSP plants and they are near the thermal stability limits of the employed carbonate-based and chloride-based salts, assuming that this temperature range is not ideal for the nitrate/nitrite-based salts as it is higher than their stability limit.

The corrosion rate produced on the L-PBF 316L ss exposed to the different molten salts was quantified using metal loss distribution, via a dimensional metrology analysis. The analyses were performed by measuring cross-section thickness of the uncorroded steel (fifty measurements in each sample) at equidistant distances before (e_0) and after (e_f) the corrosion test, as Figure 1 represents, using DMR LAS V4 light microscope (Leica, Wetzlar, Germany) with Image-Pro Plus software (version 7.0, Leica, Wetzlar, Germany). In this manner, the entire cross-section of the sample was assessed. The difference between e_0 and e_f was considered as metal loss. For the measurement of the thickness after corrosion tests (e_f), only the thickness of the uncorroded steel was considered. For this reason, it was not necessary to clean the residual salts and corrosion products from the sample's surface after the high-temperature corrosion tests. The data sets were re-ordered (from greatest to least metal loss) and plotted as metal loss (in μm) vs. cumulative probability (%). This type of plot indicates the probability (in %) of a certain degree of damage being observed.

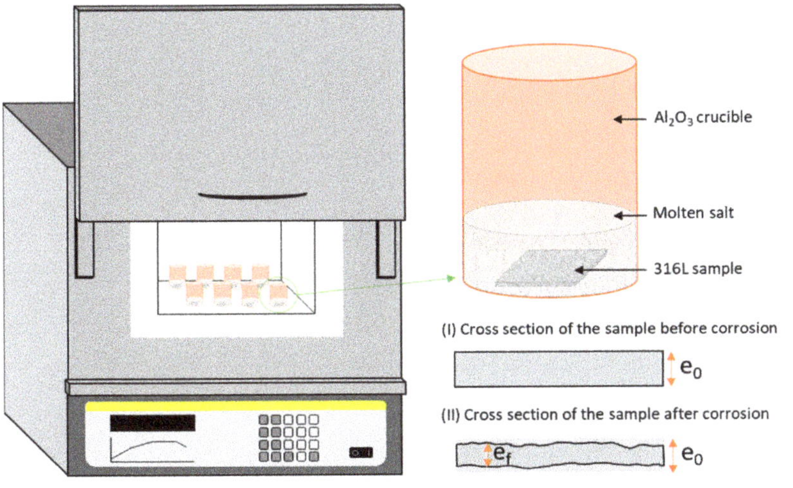

Figure 1. Scheme of the testing method employed for corrosion tests.

2.3. *Samples Characterization*

The as-built 316L austenitic stainless steel was first characterized by optical microscopy (OM) to analyze its microstructure. The steel was ground and polished and then electrochemically etched with oxalic acid to reveal the microstructure. A LEICA/DMR optical microscope equipped with Image-Pro Plus analysis software (version 7.0, Leica, Wetzlar, Germany) was employed for this purpose. The cross-sections of the sample were analyzed using a scanning electron microscope (SEM, Hitachi S-3400 N, Hitachi High-Technologies, Tokyo, Japan) equipped with a XFlash 5010 energy dispersive X-ray spectrometer (EDS, Bruker, Germany).

The X-ray diffraction (XDR) technique was used to identify the corrosion products. A Bruker AXS D8 Advance diffractometer was employed using $CuK_{\alpha 1}$ radiation (λ = 1.54 nm) in θ–2θ mode. Diffraction patterns were recorded at an angular interval of 20 to 90° and indexed using the PDF-4 database of ICDD.

3. Results and Discussion

3.1. *Characterization of the Starting Materials*

The composition of the 316L ss was 17.0 Cr, 13.1 Ni, 2.7 Mo, 1.3 Mn, 1.2 Si, and bal. Fe (in wt.%). XRD spectrum of the 316L stainless steel manufactured through L-PBF is shown in Figure 2a. The pattern shows that the sample was characterized only by the FCC austenite phase (00-003-1209, ICDD).

Figure 2b shows an optical micrograph of the main plane (ZY) of the steel. Its macroscale structure was determined by the L-PBF process, which depended on the laser-scanning pattern. In this direction, a layered microstructure can be observed, composed of a network of melt pools formed from the overlap of the laser-scanning patterns. It can also be observed that large columnar austenitic grains formed within the melt pools, typical of this manufacturing process. These were formed in the direction of the thermal gradient (Z direction) through the melt pool boundaries [37]. Figure 2c illustrates the microstructure of the alloy at higher magnification. The cellular microstructure that characterizes the melt pools is shown. This microstructure is related to the manufacturing process, where a high cooling speed induced non-equilibrium solidification conditions [38]. These cells had different morphologies, equiaxed–polygonal or elongated, depending on their growth direction.

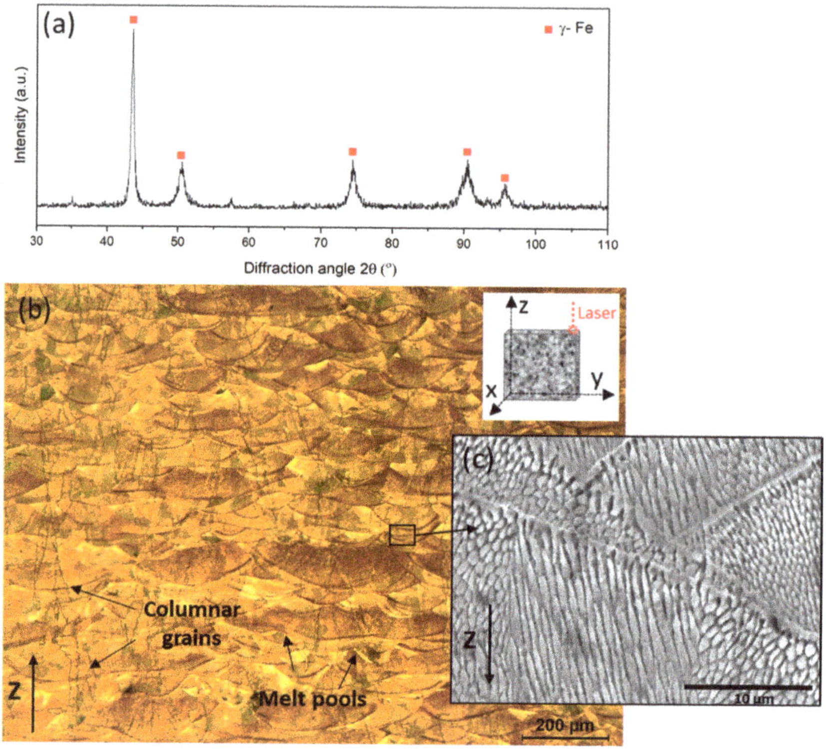

Figure 2. (**a**) X-ray diffraction analysis, (**b**) optical microscopy characterization, and (**c**) SEM micrograph of the as-built L-PBF 316L stainless steel sample.

3.2. Measurement of Metal Damage

The use of aggressive salts usually produces the growth of corrosion products, which can be lost or gained by evaporation or sublimation of gaseous species. These species can produce slight distortions in mass change measurements and, therefore, the corrosion and damage suffered by the 316L stainless steel was analyzed using dimensional metrology analysis. The metal damage produced in the 316L L-PBF stainless steel after its exposure to the different salt mixtures in terms of metal loss (in μm) vs. cumulative probability (in %) is represented in Figure 3a.

As expected, in all the tested salt mixtures, the steel suffered more damage at 700 than at 650 °C. However, the effect of temperature was negligible in the samples tested without any salt mixture, and for both temperatures, the metal loss was very low (~25 μm). Nevertheless, different trends can be observed in Figure 3a. The plots of the samples exposed to high temperatures without any molten salts (oxidation test) were straight at both temperatures and metal loss values were almost constant, indicating that no localized attack was produced, and that the oxidation process was continuous in all the samples, favoring the grown of an oxide layer with a similar thickness along the entire surface. The same trend was observed in the samples exposed to $KNaNO_x$ molten salts, probably because at these temperatures, $KNaNO_x$ is unstable and breaks down easily. However, the samples exposed to $LiNaKCO_3$ molten salts at 650 and 700 °C suffered a more heterogeneous attack. As a representative example, results for the samples tested at

700 °C showed that approximately 20% of the surface underwent more attack compared with the other 80% of the surface, which suffered less attack. The same trend was observed in the samples exposed to NaKZnCl and NaKMgCl at 700 °C, where almost 10% and 15% of sample surface, respectively, underwent more corrosion. The localized attack shown in these samples can be attributed to chloride's aggressiveness, which accelerated the corrosion process in some areas.

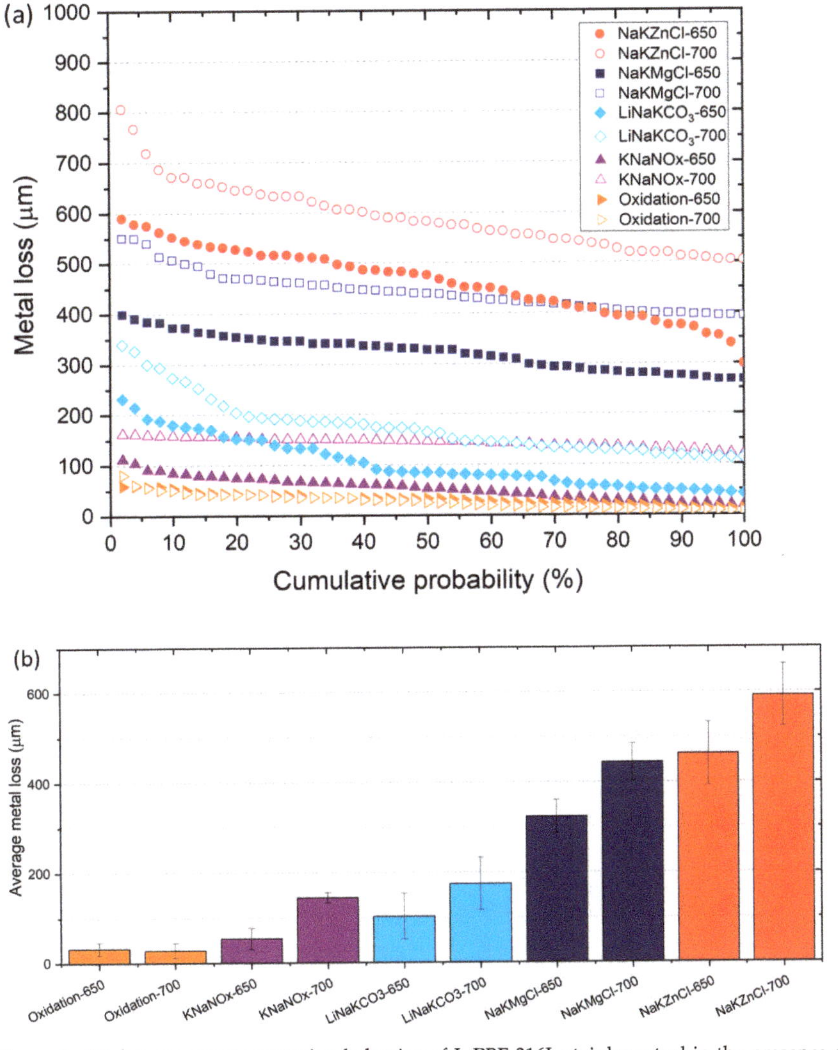

Figure 3. High-temperature corrosion behavior of L-PBF 316L stainless steel in the presence of different TES molten salts at 650 and 700 °C for 48 h: (**a**) metal loss (in μm) vs. cumulative probability (in %) and (**b**) average metal loss results (data of NaKZnCl samples from Ref [24]).

Figure 3b shows the average metal loss data reported from Figure 3a, for the L-PBF 316L stainless steel after its exposure to the different molten salts to classify these based on their aggressiveness. As expected, the most aggressive salts were those rich in chlorides, particularly NaKZnCl, followed by NaKMgCl. At 700 °C, these salts increased the metal loss twenty and fifteen times compared with the samples exposed to high temperatures without any molten salts. $KNaNO_x$ and $LiNaKCO_3$ molten salts produced lower damage in the stainless steel and increased the metal loss four and five times at 700 °C compared with the samples exposed to high temperatures without any molten salts. In addition, the metal loss increased when the exposure temperature was raised to 700 °C in all the molten salts tested.

3.3. Corrosion Products Characterization

3.3.1. Visual Analysis

The macrographs presented in Figure 4 show the surfaces of the samples after their exposure to the different TES materials. As can be seen, the composition of the salt had more influence than the exposure temperature because according to their degree of damage, there were no relevant differences between the samples tested at 650 and 700 °C. However, the exposure to the molten salts produced huge increases in oxides and corrosion product content on their surfaces. The steels exposed to high temperatures without any salts (oxidation tests) still showed their metal surface, while the samples exposed to nitrate/nitrite-based salts and carbonate-based salts no longer showed their metal surface (Figure 4). In these samples, a continuous layer of oxides/corrosion products was formed, but these still adhered to the sample. Otherwise, the 316L stainless steel samples exposed to chloride-based salts were highly attacked and corroded. The oxides and corrosion products exerted extreme delamination in these samples. These results are in accordance with those reported in Figure 3, which also confirmed not only the stronger influence of molten salt presence compared with difference in temperature in the corrosion process but also the order of aggressiveness of the studied TES materials. The nitrate/nitrite-based and carbonate-based molten salts were much less aggressive than the chloride-based molten salts.

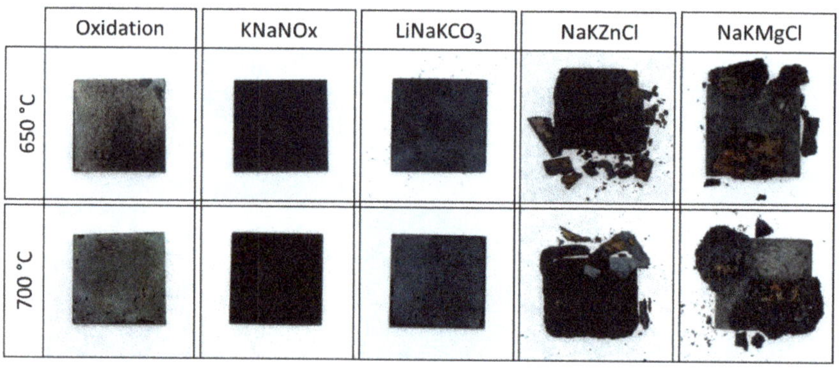

Figure 4. Surface morphology of L-PBF 316L stainless steel after its exposure to high-temperature corrosion tests at 650 and 700 °C for 48 h in the presence of different TES materials.

3.3.2. XRD Characterization

Figure 5 shows the XRD patterns of the tested samples at 650 °C, Figure 5a, and 700 °C, Figure 5b, and Table 2 shows a list of the XRD-identified species to facilitate the analysis of the results. According to the visual examination shown in Figure 4 of the samples exposed to high temperatures in the absence of molten salts, the prominent peaks were those of austenite γ-Fe; however, some small peaks of hematite (Fe_2O_3) and spinel ($(Mn,Fe)Cr_2O_4$) oxides were also identified at both tested temperatures.

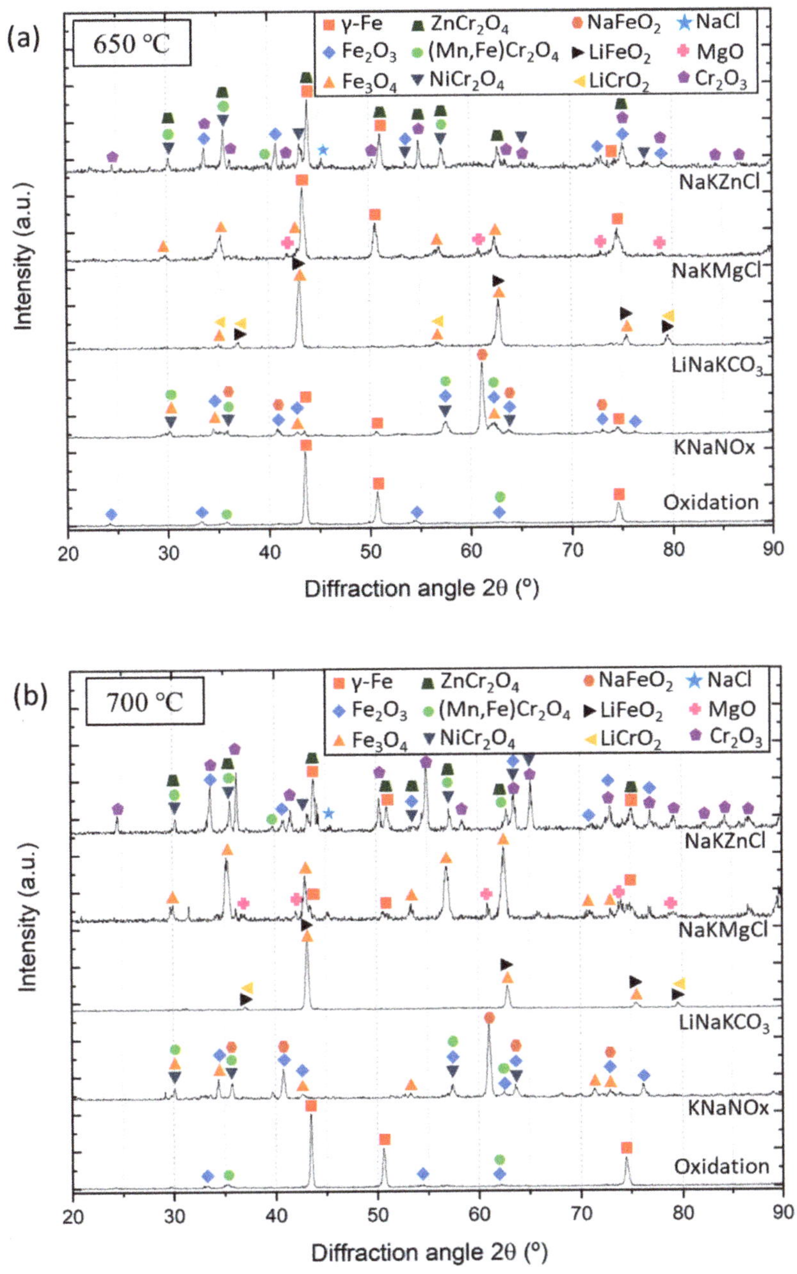

Figure 5. XRD patterns of the L-PBF 316L stainless steel after its exposure to different TES materials at (**a**) 650 and (**b**) 700 °C for 48 h.

Table 2. XRD-identified species on the L-PBF 316L stainless steel after high-temperature corrosion tests in the presence of different salt mixtures at 650 and 700 °C for 48 h. The dominant peaks are marked in bold for each test.

Steel/Deposit	Oxidation	$KNaNO_x$	$LiNaKCO_3$	NaKMgCl	NaKZnCl
650 °C	**γ-Fe** Fe_2O_3 $(Mn,Fe)Cr_2O_4$	**$NaFeO_2$** Fe_2O_3 Fe_3O_4 $NiCr_2O_4$ $(Mn,Fe)Cr_2O_4$ γ-Fe	**Fe_3O_4** **$LiFeO_2$** $LiCrO_2$	**Fe_3O_4** **MgO** γ-Fe	**Cr_2O_3** **$(Mn,Fe)Cr_2O_4$** **$NiCr_2O_4$** **$ZnCr_2O_4$** Fe_2O_3 γ-Fe NaCl
700 °C	**γ-Fe** Fe_2O_3 $(Mn,Fe)Cr_2O_4$	**$NaFeO_2$** Fe_2O_3 Fe_3O_4 $NiCr_2O_4$ $(Mn,Fe)Cr_2O_4$	**Fe_3O_4** **$LiFeO_2$** $LiCrO_2$	**Fe_3O_4** **MgO** γ-Fe	**Cr_2O_3** **$(Mn,Fe)Cr_2O_4$** **$NiCr_2O_4$** **$ZnCr_2O_4$** Fe_2O_3 γ-Fe NaCl

In the samples exposed to the $KNaNO_x$ salt mixture, the main peaks were associated with $NaFeO_2$ [39]; however, some oxides that adhered well to the surface (Figure 4), such as Fe_2O_3, Fe_3O_4, $(Mn,Fe)Cr_2O_4$, and $NiCr_2O_4$, were also identified. Peaks of γ-Fe were identified at 650 °C but not at 700 °C. This is because of the presence of a thicker layer of oxides and corrosion products in the sample tested at 700 °C (see Figure 3), which would prevent the identification of the γ-Fe phase via XRD.

The XRD patterns of the 316L stainless steel tested under $LiNaKCO_3$ molten salts at 650 and 700 °C revealed that at both temperatures, the corrosion products formed on the samples' surfaces were $LiFeO_2$, $LiCrO_2$, and Fe_3O_4. These were the main corrosion products, and Luo et al. [40] also identified these peaks when they studied the corrosion of the same stainless steel by LiNaK carbonate molten salt at 600 and 700 °C. González-Fernández et al. [41] analyzed the corrosion products formed in 310 ss (24.9% Cr) at 600 °C for a longer exposure time (600 h) in dynamic conditions and also found the presence of Fe_3O_4, $LiCrO_2$, and $Li(Fe,Ni)O_2$ as the main corrosion products. This means that the short exposure time was enough to produce the lithiation reaction of the grown oxides. González-Fernández et al. also found the presence of $MnNi_6O_8$ and K_2NiO_2; in contrast, however, in our case, these compounds were not identified at lower exposure times.

In the samples exposed to the NaKMgCl and NaKZnCl chloride-based TES materials, peaks of γ-Fe were also identified; however, in this case, the peaks appeared due to the high delamination suffered by the oxide coating, as Figure 4 previously revealed, that exposed the uncorroded material to the surface. In the samples exposed to NaKZnCl molten salt, peaks of Cr_2O_3, $ZnCr_2O_4$, and $NiCr_2O_4$ were also identified at both temperatures. However, in the NaKMgCl molten salt, the most relevant result was the presence of MgO, which other authors also identified as a corrosion product in Ni-rich alloys [42,43]. In this context, Gong et al. [44] analyzed the compatibility of Fe-based alloys with purified molten $MgCl_2$-KCl-NaCl salt at the same temperature (700 °C) for longer periods and they also found the formation of MgO as the main corrosion product after 1000 h.

3.3.3. Cross-Section Characterization

Figure 6 shows the SEM analysis of the cross-section of the L-PBF 316L ss exposed to 650 and 700 °C without any molten salts. The microstructure of the grown oxide layer was extensively characterized in a previous study [24]. It was formed of a homogeneous coating of ~6 and ~9 μm thickness on the samples tested at 650 and 700 °C, respectively, and it was composed from the inner to the top surface of $MnCr_2O_4$, Fe_2O_3, and $FeCr_2O_4$. In addition, the oxide layer adhered well to the substrate; no delamination or cracks were observed in the substrate–oxide interface, as Figure 6c shows at higher magnification for the L-PBF 316L stainless steel exposed at 700 °C for 48 h.

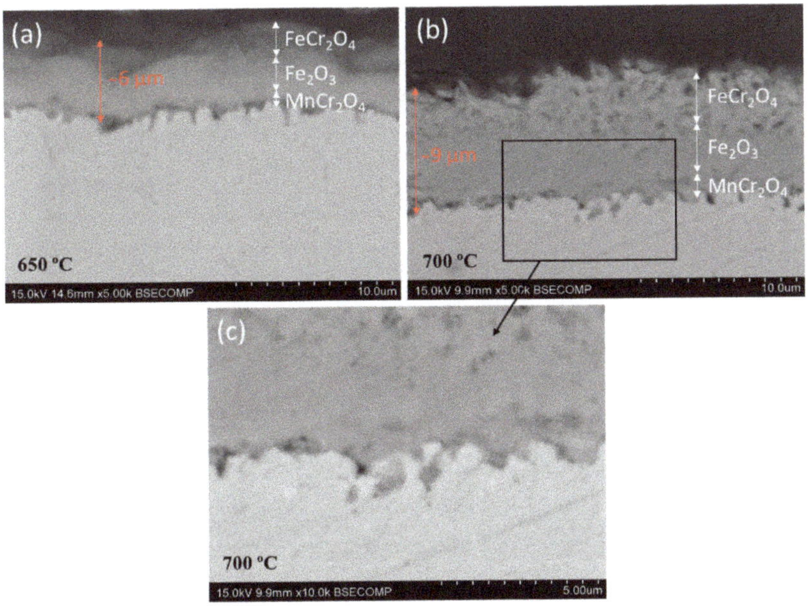

Figure 6. SEM characterization of the L-PBF 316L ss exposed to dry air atmosphere at (**a**) 650 °C and (**b**,**c**) 700 °C for 48 h at different magnification (Adapted from Ref. [24]).

Figure 7 shows the SEM micrographs of the L-PBF 316L stainless steel exposed to the $KNaNO_x$ salt at 650 °C (Figure 7a,b) and 700 °C (Figure 7c–e). The main difference observed between the two test temperatures was the thickness of the grown oxides coating, which was of ~8 µm and ~20 µm thickness at 650 and 700 °C, respectively. Figure 7a shows that there was not a great reaction of the steel with the salt mixture, probably because at this range of temperatures, $KNaNO_x$ salt is unstable and breaks down easily. Nevertheless, some important differences were seen compared with the samples tested in that absence of salt and different layers were observed in the grown coating, as the EDX map of Figure 7b shows. The inner layers were rich in Mn, Cr, and Ni and, according to the XRD results, $(Mn,Fe)Cr_2O_4$ and $NiCr_2O_4$ oxides grew. The intermediate and outer layers had more content of Fe and Na, and they may have been composed, respectively, of Fe_3O_4, Fe_2O_3, and $NaFeO_2$. The formation of iron oxides was mainly initiated by the reaction of iron with the oxygen of the atmosphere and $NaFeO_2$ was produced due to the reaction of Na_2O with these iron oxides. Ahmed [45] also found $NaFeO_2$ on 304 stainless steel after exposure to eutectic KNO_3-$NaNO_3$ and he explained the presence of a sodium iron oxide phase on the top region of the bulk corrosion scale with the following reactions (1) and (2):

$$2NaNO_2 \leftrightarrow Na_2O + NO + NO_2 \tag{1}$$

$$Na_2O + Fe_2O_3 \leftrightarrow 2NaFeO_2 \tag{2}$$

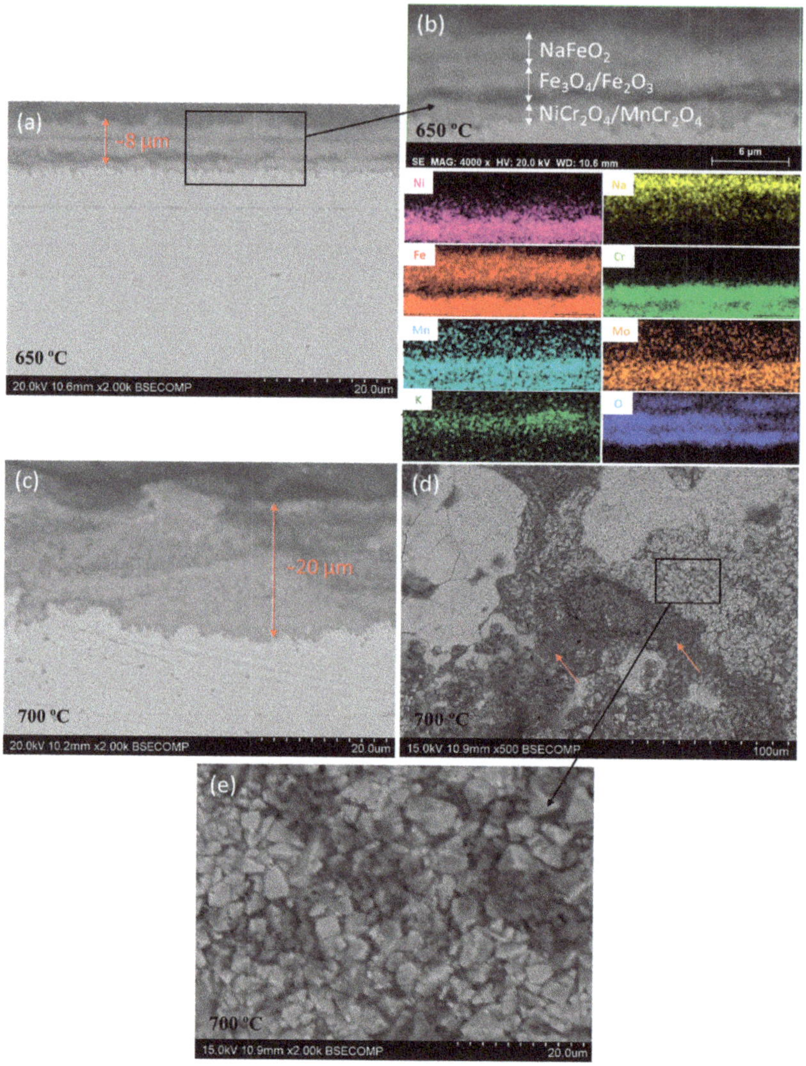

Figure 7. SEM and EDX characterization of the L-PBF 316L ss exposed to the KNaNO$_x$: (**a**,**b**) cross-sectional analysis of samples exposed at 650 °C and (**c**–**e**) cross-sectional and superficial analysis of samples exposed at 700 °C for 48 h.

In addition, the presence of K_2CrO_4, which was not identified with XRD, would explain the presence of K in the EDX map, as it has been observed in Fe-Cr alloys in presence of KCl [46]. Other authors, such as Sandoval-Amador et al. [17] and Zhang et al. [47], analyzed the corrosion of conventionally manufactured 316L ss in presence of molten $NaNO_3$-KNO_3-$NaNO_2$ and $NaNO_3$-KNO_3 at lower temperatures (450 to 550 °C for 150 h), and they identified iron and spinel oxides as the main corrosion products. Gomes et al. [48] analyzed the behavior of the same stainless steel in a mixture of 60 wt.% $NaNO_3$ and 40 wt.% KNO_3 at 550 °C, and again, they found Fe_2O_3, Fe_3O_4, and $FeCr_2O_4$ as the main scale products. In their study, the thickness of the scale layer formed on AISI 316L was 6.9 ± 2.1 μm, like that reported in the present study; however, while the exposure time was longer (3000 h), the exposure temperature was 100 °C lower.

Figure 7d shows the surface of the L-PBF 316L ss after its exposure to the $KNaNO_x$ at 700 °C. The red arrows in Figure 7d indicate the accumulation of oxides in some areas and, at higher magnification, the morphology of the oxides forming small octahedral crystals of ~5 μm can be differentiated (Figure 7e). The EDX analysis performed in this area (in Figure 7d) revealed the presence of only Fe, Na, and O, which are associated with $NaFeO_2$ and Fe-rich oxides.

Figure 8 shows the SEM micrographs of the L-PBF 316L stainless steel exposed to carbonate-based ($LiNaKCO_3$) molten salts at 650 °C (Figure 8a–c) and 700 °C (Figure 8d–g) for 48 h. The continuous oxide layer shown previously using the $KNaNO_x$ salt was not observed, but the degree of corrosion was slightly superior at 650 °C, Figure 8a. In addition, Figure 8b shows at higher magnification that the area near the surface was porous and full of cavities and holes (marked with red arrows in Figure 8a,b). These vacancies were left by the depletion of some elements, thereby forming diffusion paths [49]. The formation of these diffusion paths during the high-temperature corrosion process can be attributed to unbalanced diffusional processes known as the Kirkendall effect. In this area, the main elements of the alloy (Fe and Cr) were not present, as the EDX mappings of Figure 8c revealed. In contrast, this area was enriched in Ni and Mo. Otherwise, the presence of Li could not be detected through EDX (it is important to consider the low sensitivity of EDX to light elements), but some Li-rich phases, such as $LiFeO_2$ and $LiCrO_2$, were detected with XRD, as shown in Figure 5. The high activity of Li ions in the molten salt produced their reaction with Fe and Cr-rich oxides, forming Li-rich compounds ($LiFeO_2$ and $LiCrO_2$) following reactions (3) and (4). $LiCrO_2$ has a higher solubility than $LiFeO_2$ in the molten carbonate salt; therefore, a stable $LiFeO_2$ layer gradually formed on the top.

$$Fe_2O_3 + O^{2-} + 2Li^+ \leftrightarrow 2LiFeO_2 \quad (3)$$

$$Cr_2O_3 + O^{2-} + 2Li^+ \leftrightarrow 2LiCrO_2 \quad (4)$$

Keijzer et al. [50] analyzed the presence of Li-rich compounds on AISI 304 stainless steel, showing an outermost Fe-rich layer followed by a Cr-rich layer next to the steel. The $LiCrO_2$ layer acted as a strong diffusion barrier for metal ions on top of the oxide layer, while the $LiFeO_2$ layer above helped to protect the Cr-rich layer from chromate formation. However, at the same time, the formation of these layers produced high stress in both the metal and the oxide layers and led to cracking. This probably produced the delamination of the metal at 700 °C shown in Figure 8d,e under high magnification.

Sarvghad et al. [51] tested the same molten salts with conventional 316 stainless steel and reported that the steel did not suffer severe damage and no localized dealloying was produced. However, they tested the steel at lower temperatures (450 °C). In addition, Luo et al. [40] employed the same TES material in 316L ss. They reported that the corrosion process was mainly divided into four steps: (i) oxidation of the alloy, (ii) lithiation reaction of the oxides, (iii) formation of a dual-structured corrosion scale, and (iv) spallation of corrosion products. These steps may also be valid for the present work, but probably with changes in the kinetic reactions attributed to differences in the alloy's fine sub-granular cellular microstructure caused by the manufacturing process. Figure 8f represents the surface of the 316L exposed to $LiNaKCO_3$ at 700 °C, and the Li-rich corrosion products and the oxides forming octahedral crystals are shown at higher magnification in Figure 8g.

Figure 9a shows the SEM analysis of the L-PBF 316L ss after its corrosion by NaKMgCl molten salts at 650 °C. In this case, the steel suffered severe damage, and the affected area was much deeper than with the previously analyzed molten salts. The steel was full of corrosion pathways, and different layers of corrosion products were detected, the outer one being formed by oxides. The corrosion process followed an active oxidation mechanism in which a chlorine cycle is assumed to have been involved due to the molten salt's presence, where MgO originated from the spontaneous decomposition of MgOHCl, which was generated from the hydrolysis of hydrated $MgCl_2$ as $MgCl_2{\cdot}H_2O$, following reaction (5). Then, chlorine reacted with metal elements to produce volatile metal chlorides via reaction (6), escaping the corrosion system [52]. In this context, some alloying elements like Cr, which has lower standard electrode potential in molten chlorides compared with Fe or Ni, could not avoid the corrosion process [53]. In this system, volatiles like $CrCl_4$ and oxide precipitates like MgO are considered the main corrosion products based on the thermodynamic calculation [54].

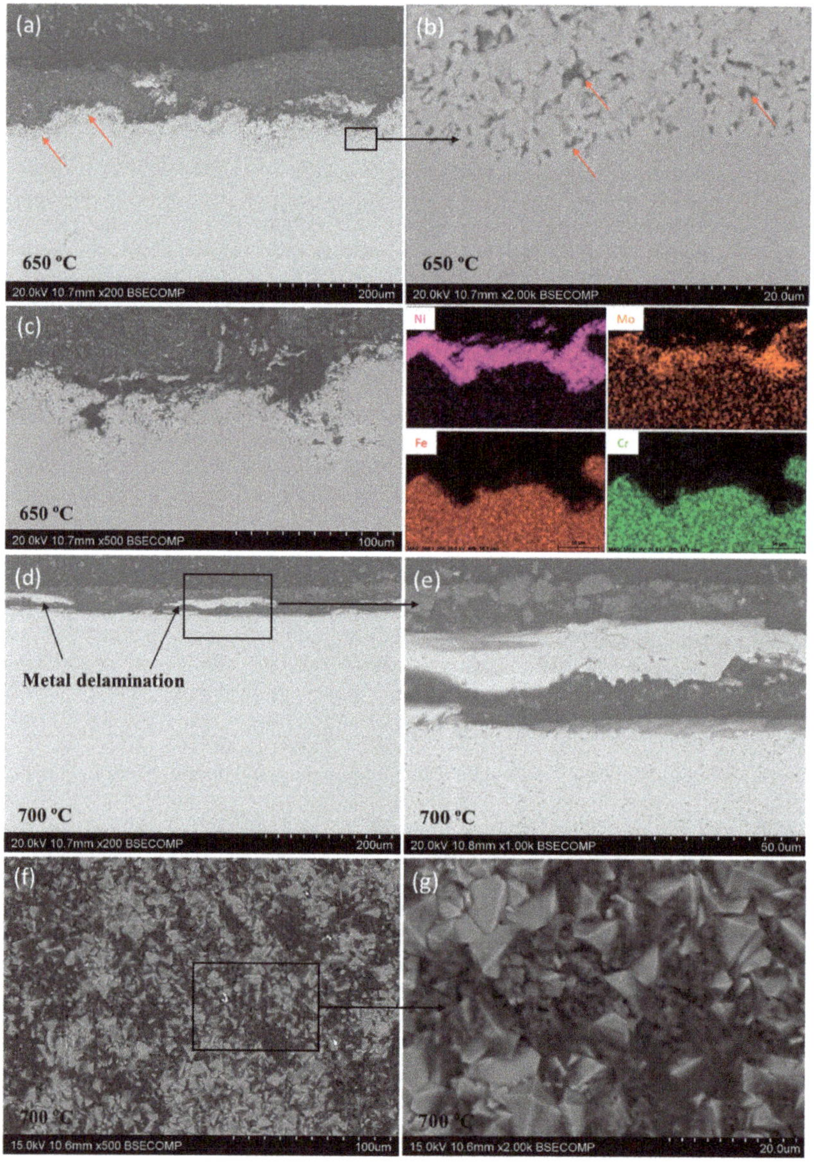

Figure 8. Cross-sectional and superficial SEM/EDX characterization of the L-PBF 316L ss after its exposure to the $LiNaKCO_3$ TES material at (**a–c**) 650 °C and (**d–g**) 700 °C for 48 h.

$$MgCl_2 + H_2O \leftrightarrow MgO + 2HCl \quad (5)$$

$$nCl_2 + 2M \leftrightarrow 2MCl_n \quad (6)$$

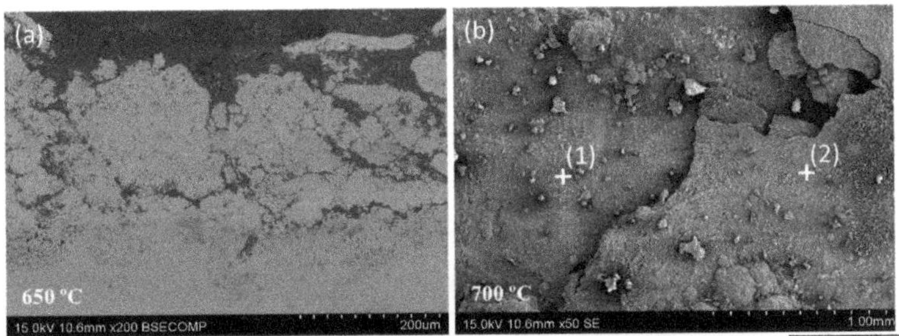

Element	weight (%)
Fe	30.9 ± 7.6
Ni	36.3 ± 8.3
Cr	2.3 ± 0.1
O	19.5 ± 9.3
Mg	-
Cl	2.8 ± 1.3
Na	7.9 ± 5.7
K	0.7 ± 0.4

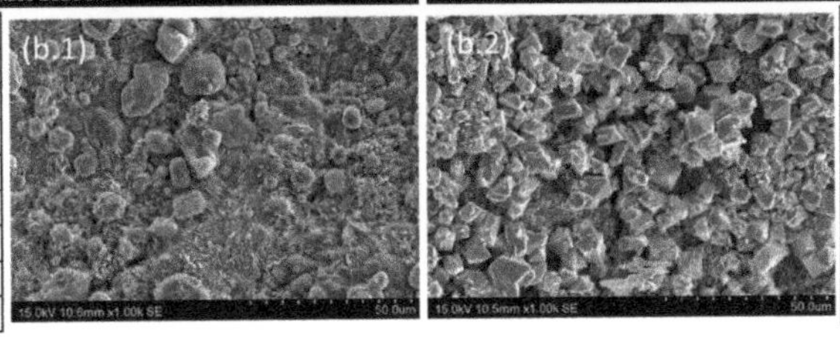

Element	weight (%)
Fe	14.7 ± 0.1
Ni	1.1 ± 0.3
Cr	44.9 ± 4.7
O	32.6 ± 2.6
Mg	6.0 ± 1.1
Cl	0.4 ± 0.1
Na	0.3 ± 0.5
K	0.2 ± 0.1

Figure 9. SEM/EDX characterization of the L-PBF 316L ss exposed to the NaKMgCl TES material for 48 h at (**a**) 650 °C and (**b**) 700 °C.

Figure 9b shows the exposed L-PBF 316L stainless steel surface at 700 °C. This micrograph shows the inner and outer layers of corrosion products because the outer layer spalled off in some regions. The border of these two different scales can be appreciated; punctual EDX analysis was performed on each region. The inner layer, marked in Figure 9b as (1), was rich in Fe (31%), Ni (36%), and O (20%); however, the presence of other elements such as Cr (2%) or Mg (no signal) was residual. In contrast, the outer layer of corrosion products, identified in Figure 9b as (2), was higher in Cr (45%), O (32%), and Mg (6%), but it had a lower amount of Fe (15%) and no evidence of Ni. According to these results, MgO, as also identified by Wang et al. [43], $MgCr_2O_4$, and other Cr-rich oxides were formed above the layer of corrosion products.

Figure 10 represents SEM micrographs of the L-PBF 316L ss after its exposure to NaKZnCl molten salts at 700 °C. The behavior of the steel under this TES material was also analyzed in a previous study [24]. Figure 10 shows how the corrosion progressed through the fine sub-grains characteristic of the L-PBF manufacturing process, without generating preferential paths for diffusion of chlorides. In this context, a high number of pathways associated with the high number of grain boundaries in the microstructure of L-PBF parts were available for the diffusion of Cr to the surface and this facilitated the enrichment in Cr near the surface and the formation of a Cr_2O_3 scale. However, the presence of the NaKZnCl molten salts induced an active oxidation mechanism, which promoted the chlorides' formation on the steel surface through the reaction of the grown Cr_2O_3 scale with the $ZnCl_2$ salt (see reaction (7)) [55] and accelerated the corrosion process through the formation of MCl_x products, as reactions (8) and (9) show. The chlorides diffused through the Fe-rich oxides to the oxide–metal interface, leading to the formation of solid iron chloride. However, the high volatility of the iron chloride provoked its diffusion backwards through the non-protective Fe-rich oxide layer.

$$ZnCl_2 + Cr_2O_3 + 1/2O_2 \leftrightarrow ZnCr_2O_4 + Cl_2 \quad (7)$$

$$Fe + Cl_2 \leftrightarrow FeCl_2 \quad (8)$$

$$Cr + Cl_2 \leftrightarrow CrCl_2 \quad (9)$$

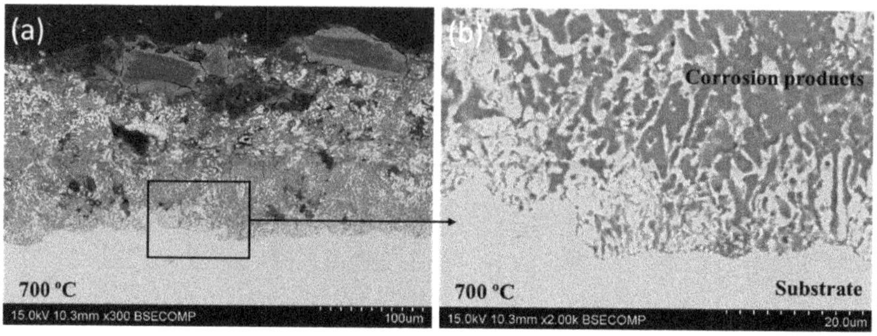

Figure 10. Cross-sectional SEM analysis of the L-PBF 316L stainless steel exposed to the NaKZnCl TES material at 700 °C for 48 h at (**a**) ×300 and (**b**) ×2000 magnification.

Once a sufficient oxygen partial pressure was reached, the chlorides destabilized and formed Cr_2O_3 or Fe_2O_3, while Cl_2 regenerated (reactions (10) and (11)). These reaction steps were continually repeated, causing severe damage in the alloy. In that case, the large amount of oxides and corrosion products on sample surface could have prevented the detection via XRD of the Fe_3O_4 phase identified in the samples exposed to NaKMgCl molten salts, but this does not mean that it was not present where Fe_2O_3 had grown on top of the Fe_3O_4 layer.

$$2FeCl_2 + 3/2O_2 \leftrightarrow Fe_2O_3 + 2Cl_2 \quad (10)$$

$$2CrCl_2 + 3/2O_2 \leftrightarrow Cr_2O_3 + 2Cl_2 \quad (11)$$

The high corrosion rates observed for the L-PBF 316L stainless steel, particularly in those samples exposed to the Cl-rich salt mixtures, highlight the importance of other alternatives, as the use of protective coatings [56,57], surface modification methods [58–60], the purification of the salts to avoid the presence of contaminants that can accelerate the corrosion process [44,61–63], or the use of corrosion inhibitors [64].

In addition, further investigations should be performed using an inert gas in absence of an air atmosphere, because hydrolysis products, e.g., HCl and MgOHCl, can accelerate the corrosion process, particularly in the presence of chloride-based salts. From this research, we can conclude that this alloy cannot be a feasible option to consider in the design of a plant which will use chloride-based molten salts at temperatures above 650 °C or, at least, not under the studied conditions where a failure in the inertization system was simulated.

4. Conclusions

The behavior of the L-PBF 316L stainless steel was more affected by the compositions of the investigated molten salts than by an increase in temperature from 650 to 700 °C, the most aggressive salts being those rich in chlorides.

A homogeneous and well-adhered oxide layer was formed on the surface of L-PBF 316L stainless steel exposed to high temperatures without any molten salts. In addition, there was the L-PBF 316L stainless steel did not react strongly with the $KNaNO_x$ salt, partially because of the degradation of the salt at the temperatures used. In these samples, the inner layers of the oxide coating were formed of $(Mn,Fe)Cr_2O_4$ and $NiCr_2O_4$ and the outer layer was composed of $NaFeO_2$, Fe_3O_4, and Fe_2O_3.

The high activity of Li ions in the $LiNaKCO_3$ molten salts produced their reaction with Fe and Cr-rich oxides, forming $LiFeO_2$ and $LiCrO_2$ as the main corrosion products, and their growth produced the depletion of Fe and Cr elements and the formation of vacancies which acted as diffusion paths in the surface of the steel.

In the samples exposed to chloride-based molten salts, the corroded area was much deeper and the corrosion process followed an active oxidation mechanism where chlorine reacted with metal elements to produce volatile metal chlorides which accelerated the corrosion process.

Author Contributions: Conceptualization, N.A.-w., S.G.-R. and M.V.U.; methodology, N.A.-w. and S.G.-R.; formal analysis, N.A.-w. and S.G.-R.; investigation, N.A.-w. and S.G.-R.; writing—original draft preparation, N.A.-w. and S.G.-R.; writing—review and editing, B.T., M.V.U. and J.R.; supervision, M.V.U. and J.R.; project administration, B.T. and J.R.; funding acquisition, B.T. and J.R. All authors have read and agreed to the published version of the manuscript.

Funding: This research was funded by Ministerio de Ciencia e Innovación (PID2021-123891OB-I00, PID2021-124341OB-C21).

Data Availability Statement: The raw/processed data required to reproduce these findings cannot be shared at this time as the data also form part of an ongoing study.

Conflicts of Interest: The authors declare no conflicts of interest.

References

1. Guccione, S.; Guedez, R. Techno-economic optimization of molten salt based CSP plants through integration of supercritical CO_2 cycles and hybridization with PV and electric heaters. *Energy* **2023**, *283*, 128528. [CrossRef]
2. Pilotti, L.; Colombari, M.; Castelli, A.F.; Binotti, M.; Giaconia, A.; Martelli, E. Simultaneous design and operational optimization of hybrid CSP-PV plants. *Appl. Energy* **2023**, *331*, 120369. [CrossRef]
3. Singh, S.P.; Gupta, D.K.; Yadav, P.; Singh, S.; Kumar, A. Influence of process parameters on capital cost, the efficiency of CSP based solar power plants—A review. *Mater. Today Proc.* **2022**, *62*, 123–130. [CrossRef]
4. Alva, G.; Liu, L.; Huang, X.; Fang, G. Thermal energy storage materials and systems for solar energy applications. *Renew. Sustain. Energy Rev.* **2017**, *68*, 693–706. [CrossRef]
5. Pelay, U.; Luo, L.; Fan, Y.; Stitou, D.; Rood, M. Thermal energy storage systems for concentrated solar power plants. *Renew. Sustain. Energy Rev.* **2017**, *79*, 82–100. [CrossRef]
6. Desideri, U.; Campana, P.E. Analysis and comparison between a concentrating solar and a photovoltaic power plant. *Appl. Energy* **2014**, *113*, 422–433. [CrossRef]
7. Ong, T.C.; Sarvghad, M.; Lippiatt, K.; Bell, S.; Will, G.; Steinberg, T.A. Investigation of the corrosion of electro-less nickel-plated alloys in molten salt and its effect on phase change properties for energy storage applications. *Sol. Energy* **2022**, *236*, 512–521. [CrossRef]
8. Li, G. Sensible heat thermal storage energy and exergy performance evaluations. *Renew. Sustain. Energy Rev.* **2016**, *53*, 897–923. [CrossRef]
9. Caraballo, A.; Galán-Casado, S.; Caballero, Á.; Serena, S. Molten salts for sensible thermal energy storage: A review and an energy performance analysis. *Energies* **2021**, *14*, 1197. [CrossRef]
10. Fernández, A.G.; Lasanta, M.I.; Pérez, F.J. Molten salt corrosion of stainless steels and low-Cr steel in CSP plants. *Oxid. Met.* **2012**, *78*, 329–348. [CrossRef]
11. Ding, W.; Bonk, A.; Bauer, T. Corrosion behavior of metallic alloys in molten chloride salts for thermal energy storage in concentrated solar power plants: A review. *Front. Chem. Sci. Eng.* **2018**, *12*, 564–576. [CrossRef]
12. de Miguel, M.T.; Lasanta, M.I.; García-Martín, G.; Díaz, R.; Pérez, F.J. Temperature effect and alloying elements impact on the corrosion behaviour of the alloys exposed to molten carbonate environments for CSP application. *Corros. Sci.* **2022**, *201*, 110274. [CrossRef]
13. Morales, M.; Gordon, S.; Fernández-Arana, Ó.; García-Marro, F.; Mateo, A.; Llanes, L.; Fargas, G. Duplex Stainless Steels for Thermal Energy Storage: Characterization of Oxide Scales Formed in Carbonate Salts at 500 °C. *Metals* **2022**, *12*, 2156. [CrossRef]
14. Liu, Q.; Xu, H.; Yin, H.; Li, N.; Wang, W.; Li, L.; Tang, Z.; Qian, Y. Corrosion behaviour of 316 stainless steel in NaCl-KCl-$MgCl_2$ salt vapour at 700 °C. *Corros. Sci.* **2022**, *194*, 109921. [CrossRef]
15. Feng, J.; Mao, L.; Yuan, G.; Zhao, Y.; Vidal, J.; Liu, L. Grain size effect on corrosion behavior of Inconel 625 film against molten MgCl2-NaCl-KCl salt. *Corros. Sci.* **2022**, *197*, 110097. [CrossRef]
16. Li, X.; Chang, L.; Liu, C.; Leng, B.; Ye, X.; Han, F.; Yang, X. Effect of thermal aging on corrosion behavior of type 316H stainless steel in molten chloride salt. *Corros. Sci.* **2021**, *191*, 109784. [CrossRef]
17. Sandoval-Amador, A.; Santander-Vega, A.J.; Amaya-Cáceres, C.C.; Estupiñán-Duran, H.A.; Peña-Ballesteros, D.Y. 316L stainless steel corrosion in molten salts $NaNO_3$ KNO_3 $NaNO_2$ simulating storage conditions. *J. Phys. Conf. Ser.* **2019**, *1159*, 012011. [CrossRef]
18. Cheng, W.J.; Chen, D.J.; Wang, C.J. High-temperature corrosion of Cr-Mo steel in molten LiNO3-NaNO3-KNO3 eutectic salt for thermal energy storage. *Sol. Energy Mater. Sol. Cells* **2015**, *132*, 563–569. [CrossRef]

19. Salarvand, V.; Sohrabpoor, H.; Mohammadi, M.A.; Nazari, M.; Raghavendra, R.; Mostafaei, A.; Brabazon, D. Microstructure and corrosion evaluation of as-built and heat-treated 316L stainless steel manufactured by laser powder bed fusion. *J. Mater. Res. Technol.* **2022**, *18*, 4104–4113. [CrossRef]
20. Brittan, A.; Mahaffey, J.; Adam, D.; Anderson, M. Mechanical and Corrosion Response of 316SS in Supercritical CO_2. *Oxid. Met.* **2021**, *95*, 409–425. [CrossRef]
21. Li, H.; Yang, X.; Yin, X.; Wang, X.; Tang, J.; Gong, J. Effect of Chloride Impurity on Corrosion Kinetics of Stainless Steels in Molten Solar Salt for CSP Application: Experiments and Modeling. *Oxid. Met.* **2021**, *95*, 311–332. [CrossRef]
22. Li, H.; Feng, X.; Wang, X.; Yang, X.; Tang, J.; Gong, J. Impact of temperature on corrosion behavior of austenitic stainless steels in solar salt for CSP application: An electrochemical study. *Sol. Energy Mater. Sol. Cells* **2022**, *239*, 111661. [CrossRef]
23. Yin, Y.; Rumman, R.; Chambers, B.A.; Liu, M.; Jacob, R.; Belusko, M.; Bruno, F.; Lewis, D.A.; Andersson, G.G. Corrosion interface formation in thermally cycled stainless steel 316 with high-temperature phase change material. *Sol. Energy Mater. Sol. Cells* **2021**, *225*, 111062. [CrossRef]
24. Abu-warda, N.; García-Rodríguez, S.; Torres, B.; Utrilla, M.V.; Rams, J. Impact of the additive manufacturing process on the high-temperature corrosion of 316L steel in the presence of NaCl-KCl $ZnCl_2$ molten solar salt. *J. Mater. Res. Technol.* **2022**, *20*, 3949–3961. [CrossRef]
25. Herzog, D.; Seyda, V.; Wycisk, E.; Emmelmann, C. Additive manufacturing of metals. *Acta Mater.* **2016**, *117*, 371–392. [CrossRef]
26. He, X.; He, B.B. Microstructural Evolution and Mechanical Behavior of Additively Manufactured 316 L Subjected to Varied Deformation Directions. *Met. Mater. Int.* **2024**, *30*, 25–38. [CrossRef]
27. Kumaran, M.; Senthilkumar, V. Influence of Heat Treatment on Stainless Steel 316L Alloy Manufactured by Hybrid Additive Manufacturing Using Powder Bed Fusion and Directed Energy Deposition. *Met. Mater. Int.* **2023**, *29*, 467–484. [CrossRef]
28. Khaleghifar, F.; Razeghi, K.; Heidarzadeh, A.; Mousavian, R.T. Effect of hydrogen on the tensile behavior of austenitic stainless steels 316l produced by laser-powder bed fusion. *Metals* **2021**, *11*, 586. [CrossRef]
29. Baumers, M.; Dickens, P.; Tuck, C.; Hague, R. The cost of additive manufacturing: Machine productivity, economies of scale and technology-push. *Technol. Forecast. Soc. Chang.* **2016**, *102*, 193–201. [CrossRef]
30. Siri, C.; Popa, I.; Vion, A.; Langlade, C.; Chevalier, S. Impact of Selective Laser Melting Additive Manufacturing on the High Temperature Behavior of AISI 316L Austenitic Stainless Steel. *Oxid. Met.* **2020**, *94*, 527–548. [CrossRef]
31. Gerstler, W.D.; Barua, A.; Huang, S.; Erno, D.J.; Wang, Y.; Ju, S.; Thiagarajan, N. Binder Jet Additive Manufacturing Process and Material Characterization for High Temperature Heat Exchangers Used in Concentrated Solar Power Applications. *Metals* **2023**, *13*, 617. [CrossRef]
32. Spiegel, M.; Schraven, P. Corrosion of Austenitic Steels and Nickel Alloys in Molten KNO_3–$NaNO_3$ at Different Temperatures: Role of Alloying Elements. *Oxid. Met.* **2021**, *96*, 145–155. [CrossRef]
33. Morales, M.; Cabezas, L.; Castro-Alloca, M.; Fargas, G.; Llanes, L.; Mateo, A. Corrosion Evaluation of Austenitic and Duplex Stainless Steels in Molten Carbonate Salts at 600 °C for Thermal Energy Storage. *Metals* **2022**, *12*, 2190. [CrossRef]
34. Du, L.; Ding, J.; Tian, H.; Wang, W.; Wei, X.; Song, M. Thermal properties and thermal stability of the ternary eutectic salt NaCl-CaCl2-MgCl2 used in high-temperature thermal energy storage process. *Appl. Energy* **2017**, *204*, 1225–1230. [CrossRef]
35. Li, Y.; Xu, X.; Wang, X.; Li, P.; Hao, Q.; Xiao, B. Survey and evaluation of equations for thermophysical properties of binary/ternary eutectic salts from NaCl, KCl, $MgCl_2$, $CaCl_2$, $ZnCl_2$ for heat transfer and thermal storage fluids in CSP. *Sol. Energy* **2017**, *152*, 57–79. [CrossRef]
36. *ISO/FDIS 17224:2014*; Corrosion of Metals and Alloys—Test Method for High Temperature Corrosion Testing of Metallic Materials by Application of a Deposit of Salt, Ash, or Other Substances. ISO: Geneva, Switzerland, 2014.
37. Liverani, E.; Toschi, S.; Ceschini, L.; Fortunato, A. Effect of selective laser melting (SLM) process parameters on microstructure and mechanical properties of 316L austenitic stainless steel. *J. Mater. Process. Technol.* **2017**, *249*, 255–263. [CrossRef]
38. Wang, D.; Song, C.; Yang, Y.; Bai, Y. Investigation of crystal growth mechanism during selective laser melting and mechanical property characterization of 316L stainless steel parts. *Mater. Des.* **2016**, *100*, 291–299. [CrossRef]
39. Abu-Warda, N.; López, A.J.; Pedraza, F.; Utrilla, M.V. Corrosion behavior of T24, T92, VM12, and AISI 304 steels exposed to KCl–NaCl–K_2SO_4–Na_2SO_4 salt mixtures. *Mater. Corros.* **2021**, *72*, 936–950. [CrossRef]
40. Luo, J.; Deng, C.K.; ul Haq Tariq, N.; Li, N.; Han, R.F.; Liu, H.H.; Wang, J.Q.; Cui, X.Y.; Xiong, T.Y. Corrosion behavior of SS316L in ternary Li_2CO_3–Na_2CO_3–K_2CO_3 eutectic mixture salt for concentrated solar power plants. *Sol. Energy Mater. Sol. Cells* **2020**, *217*, 110679. [CrossRef]
41. González-Fernández, L.; Intxaurtieta-Carcedo, M.; Bondarchuk, O.; Grosu, Y. Effect of dynamic conditions on high-temperature corrosion of ternary carbonate salt for thermal energy storage applications. *Sol. Energy Mater. Sol. Cells* **2022**, *240*, 111666. [CrossRef]
42. Wang, J.W.; Bao, Z.; Xiang, R.Q.; Hu, G.H.; Hu, Y.K.; Ma, Q.; Li, B.; Jia, H.G. Corrosion behavior of Fe-37 wt% Ni in molten NaCl-$MgCl_2$ and the effects of alloy element Sc on its corrosion resistance. *Mater. Chem. Phys.* **2020**, *243*, 122392. [CrossRef]
43. Wang, J.; Zhang, C.; Li, Z.; Zhou, H.; He, J.; Yu, J. Corrosion behavior of nickel-based superalloys in thermal storage medium of molten eutectic NaCl-MgCl2 in atmosphere. *Sol. Energy Mater. Sol. Cells* **2017**, *164*, 146–155. [CrossRef]
44. Gong, Q.; Shi, H.; Chai, Y.; Yu, R.; Weisenburger, A.; Wang, D.; Bonk, A.; Bauer, T.; Ding, W. Molten chloride salt technology for next-generation CSP plants: Compatibility of Fe-based alloys with purified molten $MgCl_2$-KCl-NaCl salt at 700 °C. *Appl. Energy* **2022**, *324*, 119708. [CrossRef]

45. Ahmed, O. Corrosion Behaviour of AISI 304 Stainless Steel in Contact with Eutectic Salt for Concentrated Solar Power Plant Applications. Master's Thesis, Materials Science Engineering, College of Engineering and Computer Science, University of Central Florida, Orlando, FL, USA, 2013.
46. Li, Y.S.; Niu, Y.; Spiegel, M. High temperature interaction of Al/Si-modified Fe–Cr alloys with KCl. *Corros. Sci.* **2007**, *49*, 1799–1815. [CrossRef]
47. Zhang, X.; Zhang, C.; Wu, Y.; Lu, Y. Experimental research of high temperature dynamic corrosion characteristic of stainless steels in nitrate eutectic molten salt. *Sol. Energy* **2020**, *209*, 618–627. [CrossRef]
48. Gomes, A.; Navas, M.; Uranga, N.; Paiva, T.; Figueira, I.; Diamantino, T.C. High-temperature corrosion performance of austenitic stainless steels type AISI 316L and AISI 321H, in molten Solar Salt. *Sol. Energy* **2019**, *177*, 408–419. [CrossRef]
49. Wang, C.; Liu, Z.; Chen, W.; Chen, D. Effect of metallic ion products on the corrosion of GH3535 alloy in a eutectic (Li, Na, K) F melt. *J. Mater. Res. Technol.* **2023**, *22*, 1014–1025. [CrossRef]
50. Keijzer, M.; Lindbergh, G.; Hemmes, K.; Van der Put, P.J.J.M.; Schoonman, J.; De Wit, J.H.W. Corrosion of 304 stainless steel in molten carbonate fuel cells. *J. Electrochem. Soc.* **1999**, *146*, 2508–2516. [CrossRef]
51. Sarvghad, M.; Steinberg, T.A.; Will, G. Corrosion of stainless steel 316 in eutectic molten salts for thermal energy storage. *Sol. Energy* **2018**, *172*, 198–203. [CrossRef]
52. Junwei, W.; Rongjiang, Z.; Siyu, C. Corrosion Behavior and Mechanism of Fe-14Cr-Mn Alloy in Molten Eutectic NaCl-$MgCl_2$ at Different Temperatures. *Rare Met. Mater. Eng.* **2017**, *46*, 935–941. [CrossRef]
53. Plambeck, J.A. Electromotive force series in molten salts. *J. Chem. Eng. Data* **1967**, *12*, 77–82. [CrossRef]
54. Liu, B.; Wei, X.; Wang, W.; Lu, J.; Ding, J. Corrosion behavior of Ni-based alloys in molten NaCl-$CaCl_2$-$MgCl_2$ eutectic salt for concentrating solar power. *Sol. Energy Mater. Sol. Cells* **2017**, *170*, 77–86. [CrossRef]
55. Wang, M.; Zeng, S.; Zhang, H.; Zhu, M.; Lei, C.; Li, B. Corrosion behaviors of 316 stainless steel and Inconel 625 alloy in chloride molten salts for solar energy storage. *High Temp. Mater. Process.* **2020**, *39*, 340–350. [CrossRef]
56. Grosu, Y.; Anagnostopoulos, A.; Navarro, M.E.; Ding, Y.; Faik, A. Inhibiting hot corrosion of molten Li_2CO_3-Na_2CO_3-K_2CO_3 salt through graphitization of construction materials for concentrated solar power. *Sol. Energy Mater. Sol. Cells* **2020**, *215*, 110650. [CrossRef]
57. Luo, J.; ul Haq Tariq, N.; Liu, H.H.; Li, N.; Zhao, L.J.; Cui, X.Y.; Xiong, T.Y. Robust corrosion performance of cold sprayed aluminide coating in ternary molten carbonate salt for concentrated solar power plants. *Sol. Energy Mater. Sol. Cells* **2022**, *237*, 111573. [CrossRef]
58. González-Fernández, L.; Anagnostopoulos, A.; Karkantonis, T.; Bondarchuk, O.; Dimov, S.; Chorążewski, M.; Ding, Y.; Grosu, Y. Laser-induced carbonization of stainless steel as a corrosion mitigation strategy for high-temperature molten salts applications. *J. Energy Storage* **2022**, *56*, 105972. [CrossRef]
59. González-Fernández, L.; Anagnostopoulos, A.; Karkantonis, T.; Dimov, S.; Chorążewski, M.; Ding, Y.; Grosu, Y. Laser-texturing of stainless steel as a corrosion mitigation strategy for high-temperature molten salts applications under dynamic conditions. *Sol. Energy Mater. Sol. Cells* **2023**, *257*, 112380. [CrossRef]
60. Encinas-Sánchez, V.; Batuecas, E.; Macías-García, A.; Mayo, C.; Díaz, R.; Pérez, F.J. Corrosion resistance of protective coatings against molten nitrate salts for thermal energy storage and their environmental impact in CSP technology. *Sol. Energy* **2018**, *176*, 688–697. [CrossRef]
61. Ong, T.-C.; Sarvghad, M.; Lippiatt, K.; Griggs, L.; Ryan, H.; Will, G.; Steinberg, T.A. Review of the solubility, monitoring, and purification of impurities in molten salts for energy storage in concentrated solar power plants. *Renew. Sustain. Energy Rev.* **2020**, *131*, 110006. [CrossRef]
62. Ding, W.; Yang, F.; Bonk, A.; Bauer, T. Molten chloride salts for high-temperature thermal energy storage: Continuous electrolytic salt purification with two Mg-electrodes and alternating voltage for corrosion control. *Sol. Energy Mater. Sol. Cells* **2021**, *223*, 110979. [CrossRef]
63. Ding, W.; Gomez-Vidal, J.; Bonk, A.; Bauer, T. Molten chloride salts for next generation CSP plants: Electrolytical salt purification for reducing corrosive impurity level. *Sol. Energy Mater. Sol. Cells* **2019**, *199*, 8–15. [CrossRef]
64. Ding, W.; Shi, H.; Jianu, A.; Xiu, Y.; Bonk, A.; Weisenburger, A.; Bauer, T. Molten chloride salts for next generation concentrated solar power plants: Mitigation strategies against corrosion of structural materials. *Sol. Energy Mater. Sol. Cells* **2019**, *193*, 298–313. [CrossRef]

metals

MDPI

Article

Unlocking the Potential of Sebacate: Investigating Its Role in the Inhibition of Filiform Corrosion on Organic Coated Steel

Andrea Cristoforetti *, Stefano Rossi, Flavio Deflorian and Michele Fedel

Department of Industrial Engineering, University of Trento, Via Sommarive n. 9, 38123 Trento, Italy; stefano.rossi@unitn.it (S.R.); flavio.deflorian@unitn.it (F.D.); michele.fedel@unitn.it (M.F.)
* Correspondence: andrea.cristoforetti@unitn.it; Tel.: +39-0461-282403

Abstract: The study investigated the effect of sebacate as a corrosion inhibitor for acrylic-coated steel. Specifically, it examined its impact on mitigating a frequent case of paint delamination, known as filiform corrosion (FFC), through a chosen weathering test designed to stress the degradation of the produced samples. Sebacate was demonstrated to be an efficient organic molecule for enhancing the corrosion resistance of steel. This efficacy was evaluated through electrochemical characterization based on electrochemical impedance spectroscopy measurements and potentiodynamic polarization curves, including the application of an FFC susceptibility prediction methodology based on measurements obtained in FFC-simulated electrolytes. An inhibition efficiency of 98% was measured in near-neutral saline solutions compared to conditions lacking inhibitor presence. During FFC simulation, the primary effect observed was associated with a reduction in cathodic activity evolution. Furthermore, a significant reduction in corrosion creep evolution of 35% was found. These experimental findings aligned closely with the outcomes projected by the simulated investigations.

Keywords: steel; corrosion; corrosion inhibitors; organic coatings; sebacate; filiform corrosion; weathering; material degradation

1. Introduction

Organic coatings are widely employed to safeguard steel substrates from corrosion while preserving their mechanical and functional integrity [1–4]. These coatings typically need to fulfill three primary requirements: adhesion to the substrate, barrier properties against ions and water, and corrosion inhibition [5]. Traditional corrosion inhibitors, often containing chromium (VI) and other toxic compounds, pose environmental and health concerns, necessitating the exploration of safer alternatives [6,7]. Over the years, various environmentally friendly inhibitors have been investigated [8–11], including carboxylic-acid-based molecules known for their ability to form complexes with iron ions, thereby competing effectively with chloride ions [12,13]. Despite their efficacy, the precise mechanism of corrosion mitigation by carboxylates and dicarboxylates remains incompletely understood, although the hydrophobic nature of the resulting thin layer is believed to play a crucial role [14–17]. Notably, the efficiency of carboxylates is influenced by their chain length, with a critical length observed for both mono- and di-carboxylates beyond which their inhibition effectiveness drops significantly due to micelle formation [13,18]. For dicarboxylates on mild steel, there is typically a minor rise in inhibition with increasing chain lengths to a moderate extent. This is succeeded by a rapid escalation to high levels of inhibition, which then experiences a sudden decrease at longer chain lengths [18]. This intricate behavior reflects the simultaneous occurrence of competing processes, including the formation of carboxylate complexes, the adsorption of molecules, and the formation of micelles or hemi-micelles. Hefter et al. [18] depicted a model of the surface interaction mechanism, suggesting that the interplay of these equilibria, potentially complicated further by their kinetics, accounts for the observed effects as chain lengths vary.

Moreover, a correlation was established between corrosion inhibition and the number of carboxylate groups [19]. This study revealed the dependency of the corrosion efficiency of organic molecules containing carboxylate groups on molecular structure features, such as steric effects, electrostatic effects, and electron attractor or donor capabilities.

Carboxylates are classified as non-oxidizing inhibitors, offering corrosion protection in aerated solutions; on the other hand, they demonstrate ineffectiveness in de-aerated solutions [20–22]. This behavior, when dissolved oxygen is present, could be explained by the conversion of anodically generated soluble Fe(II) compounds into the more stable and protective ferric state (Fe(III)). Carboxylates primarily target localized defects within the thin oxide layer, whereby corrosion is mitigated by sealing these defects with corresponding weakly soluble Fe(III) compounds and molecule adsorption in the substrate [23,24]. In this context, a dicarboxylate compound such as disodium sebacate (SB, $C_{10}H_{16}Na_2O_4$), shown in Figure 1, is a suitable candidate to consider. Amidst the pursuit of environmentally benign inhibitors, in recent studies, SB has emerged particularly due to its potential for intercalation in the burgeoning field of a novel type of ceramic-based drug delivery system, namely layered double hydroxides, offering promise as an efficient smart pigment [5,13,25,26]. Thanks to the ion exchange properties inherent in such a lamellar ceramic structure, molecules such as corrosion inhibitors, once intercalated during synthesis or in a subsequent treatment phase, can be accommodated within the structure and gradually released in the presence of other ions, typically chlorides [27]. However, their effectiveness in mitigating the specific yet common paint failure known as filiform corrosion (FFC) [28] remains largely unexplored. This lacuna persists even though the delamination mechanism underlying FFC stems from a combination of distinct reactions and chemical environments at the metal–paint interface [29,30]. Consequently, a targeted investigation is warranted, as the performance metrics derived from cathodic delamination tests, adhesion assessments, and water permeation studies may not be directly applicable to this particular case study. FFC progresses through an anodic undermining process at the metal–paint interface, driven by differential aeration cell formation along the filament's length [31]. This phenomenon is influenced by factors such as humidity levels, coating defects, intermetallic particles, and oxygen availability [32,33]. Conceptually, FFC can be viewed as an electrochemical process characterized by spatially distinct anodic and cathodic regions along the threads [34,35]. The progression of FFC is driven by a potential difference (ΔE_{FFC}), with current density ($i_{CORR\,FFC}$) indicative of its propagation. While FFC usually does not affect structural integrity, it causes severe paint detachment. This compromises the protective coating and can lead to further general corrosion.

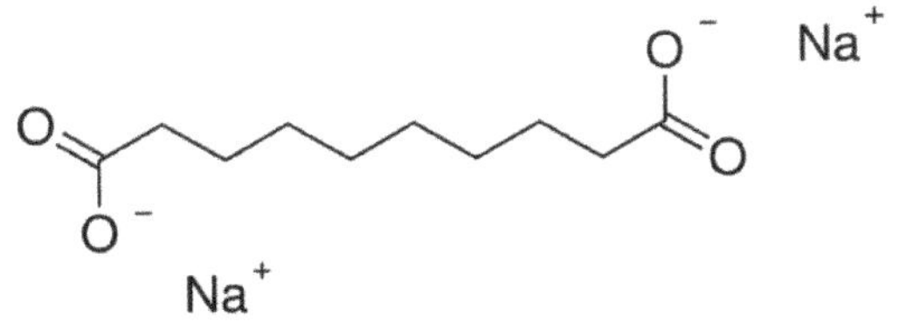

Figure 1. Representation of disodium SB structure.

This study seeks to explore the effectiveness of SB as a corrosion inhibitor for preventing FFC in acrylic-coated steel substrates. Initially, a comprehensive inhibitor evaluation involving chemical and electrochemical analyses, such as Fourier transform infrared spectroscopy (FTIR-ATR), potentiodynamic polarization curves (PDPs), and electrochemical impedance spectroscopy (EIS), was conducted. Subsequently, a detailed investigation into preventing anodic undermining failure was undertaken. The necessity of these assessments arises from the need to comprehensively understand the efficacy of SB in corrosion prevention, particularly in the context of FFC on organic-coated steel, which has remained understudied. Additionally, this study introduces a novel electrochemical simulation method to evaluate the inhibition efficiency specifically for FFC, with the results validated by observing actual corrosion creep development during humidostatic weathering tests. The corrosion mechanisms can vary notably between acidic and neutral aqueous environments, largely due to the presence or distinct properties of metal oxide/hydroxide films at elevated pH levels. Consequently, extrapolating conclusions regarding the probable effectiveness of inhibitors in neutral solutions from data acquired in acidic solutions can be misleading or even lead to adverse outcomes.

2. Materials and Methods

2.1. SB Electrochemical Characterization in a Near-Neutral Environment

In the electrochemical characterization of SB, bare R36 Q-Panel Steel A1008 surfaces (composition: C max. 0.15 wt.%; Mg max. 0.6 wt.%; P max. 0.03 wt.%; S max. 0.035 wt.%; and Fe. Bal.) were pickled for 15 min in 2 M HCl, followed by immersion in a 0.5 M NaCl solution containing 3 g/L of SB. An Autolab 302N potentiostat/galvanostat/FRA (Metrohm AG, Herisau, Switzerland) was employed with a three-electrode setup, with the steel panel serving as the working electrode, while platinum and Ag/AgCl/3.5 M KCl electrodes were utilized as the counter and reference electrodes, respectively. EIS spectra were recorded at the open circuit potential (OCP) over a frequency range of 100 kHz to 0.01 Hz, with a signal amplitude of 5 mV (RMS) and 6 points per decade, after 3 h of immersion. The obtained EIS data were analyzed using ZSIMPWIN® 3.50 software to derive equivalent electric circuits (EECs). The PDPs were obtained at a scan rate of 0.2 mV/s, ranging from −50 mV to +700 mV vs. the OCP, following a 3 h wait period.

2.2. Electrochemical Simulated Study on the Inhibition of FFC

Furthermore, to investigate the protective effects of the inhibitor against FFC, PDPs were also collected employing an anolyte/catholyte simulating setup (as described in refs. [5,35]). Also, in this case, 3 g/L of SB powder was introduced into the testing electrolytes. The anodic and cathodic branches of the PDPs were individually collected following a 3 h waiting period, during which the OCP was monitored (Appendix A, Figure A1). The anodic and cathodic branches were acquired through two separate tests by polarizing the system relative to the OCP. The anodic branch ranged from −50 mV to +300 mV, while the cathodic branch spanned from +50 mV to −300 mV. According to the procedure adopted in references [5,35], the anolyte contains 0.1 M $FeCl_3$, as it is expected that some iron dissolution will occur at the anodic site. The pH was adjusted to 1.5 by gradually adding 33% HCl. Recognizing that at the anodic site, the oxygen levels were likely reduced due to limited availability derived from the directionality of the supply through the filaments and the high ionic concentration [29,31], nitrogen was continuously purged into the anolyte to decrease the dissolved oxygen to around 1.4 ppm, measured using a PCE-PHD 1 probe (PCE Instruments, Capannori (LU), Italy). Considering that the cathodic reaction induces localized alkalinization due to the oxygen reduction reaction on the steel substrate ($4Fe^{2+} + O_2 + 2H_2O \rightarrow 4Fe^{3+} + 4OH^-$) [32,33], the catholyte was prepared using a NaOH aqueous solution with a pH of 12 [29,32]. This arrangement was anticipated to promote the passivation of the steel surface [36]. Continuous air bubbling was employed to maintain stable dissolved oxygen levels in the electrolyte.

2.3. Preparation of Acrylic-Coated Steel Samples

The effectiveness of the protective action of the SB was tested on coated steel samples composed of steel panels painted with two layers of a bicomponent polyamide-based clearcoat (supplied by Palini Vernici, Pisogne, BG, Italy). The primer layer was loaded with SB at a concentration of 1 wt.%. The top coat was a simple paint-covering layer (the same base composition as the primer) to avoid possible leaching of the particles toward the outer side of the coating. The steel substrate was initially cleaned by degreasing it with acetone under ultrasonic treatment. Subsequently, the surface was pickled in a 2 M HCl acid solution for a duration of 15 min. Paint was applied using the automatic applier Elcometer 4340 (Manchester, UK) to deposit 100 μm of wet thickness for each layer. Each layer applied underwent curing at 60 °C for 1 h, reaching a total dry thickness of 75 μm. To promote uniform dispersion of the pigments and prevent agglomeration after mixing them into the liquid paint, an ultrasonication step was conducted for a period of 10 min.

2.4. Characterization and Aging of Acrylic-Coated Steel Samples

Three painted samples of each type were aged following ASTM standard 2803 [37] to observe the spread of the FFC. A 45 mm long and 1 mm wide scratch was created on each specimen. The coated panels were protected with adhesive tape to avoid early cut-edge failure. Initially, a 5 h contamination stage in a neutral salt spray chamber [38] was conducted before subjecting the specimens to 750 h of humidostatic aging. During the aging process, the temperature was maintained at a constant 40 °C, and the relative humidity was set at 80%.

Similar coated plates were immersed in a NaCl solution (0.5 M) for 500 h. The coating system was monitored periodically by recording the EIS response. Spectra were recorded at room temperature (25 °C) in an aqueous solution with a quasi-neutral pH (6.2) and with 7.5 ppm dissolved oxygen, measured using a PCE-PHD 1 probe (PCE Instruments, Capannori (LU), Italy). The spectra were recorded at a voltage amplitude of ±15 mV with respect to the open circuit potential in a frequency range from 0.01 Hz to 100 kHz. A built-in-house mobile cell with a platinum counter electrode with an area of 0.75 cm^2 and an Ag/AgCl/3.5 M KCl reference electrode with a planar circular test area of 9 cm^2 was used as the working electrode. EIS data collection was performed to evaluate the corrosion protection properties of the paint during the development of the degradation of the coating system and the activation of the pigments loaded in the primer layer; each sample was analyzed periodically.

3. Results

3.1. SB Effect Characterization

From an electrochemical point of view, the SB's behavior in near-neutral conditions highlighted enhanced corrosion resistance, both in the PDPs and EIS. Comparing the outcomes with a reference case of blank saline solution (Ref.), SB's adsorption, with the formation of insoluble compounds on the surface [23,24], provided notably higher corrosion potentials and decreased the current densities by two orders of magnitude (Figure 2a) when the sample was anodically polarized. The difference in the OCP is comparable in the case of the cathodic branches of the PDPs, although concerning the cathodic activity current density, the inhibitor showed similar values to the reference curve (Figure 2b). On average, the steel substrate undergoing interaction with the SB experienced an increase in its OCP of +0.46 ± 0.09 V. These results are consistent with the information reported in the literature regarding specific cases of using SB as an inhibitor for steel, confirming its classification as an anodic inhibitor when tested in a nearly neutral aqueous environment [13,26].

On the other hand, the impedance modulus at low frequencies increased by more than one order of magnitude (Figure 2c). The changes at the metal interface were evaluated according to EEC analysis utilizing an R(QR) circuit, where Q stands for a constant-phase element (CPE) suitable for taking into account the deviations from the ideal capacitive behavior of the electrical double layer. Its definition, according to Equation (1), involves the parameters of the pre-exponential Y_0 and the exponent n, the values obtained for which are reported in Table 1. Moreover, the remaining parameters extracted were attributed to the electrolyte and the polarization resistance (R_s and R_{ct}), from which an index representative of the inhibition efficiency could be calculated [26,39] following Equation (2). Thanks to the higher magnitude of R_{ct} when SB was involved, the inhibition stood at around 98% in this case study and placed SB as one of the most successful inhibitors within the field of organic compounds [26,40]. In conjunction with a higher impedance modulus at low frequencies, SB's presence significantly impacts the EIS spectrum by broadening the time constant of the phase angle at medium to low frequencies, testifying to changes in the structure of the double layer at the metal interface.

$$Q(\omega) = \left[Y_0(j\omega)^n\right]^{-1}, \tag{1}$$

$$IE\% = (R_{ct} - R_{ct0})/R_{ct} \tag{2}$$

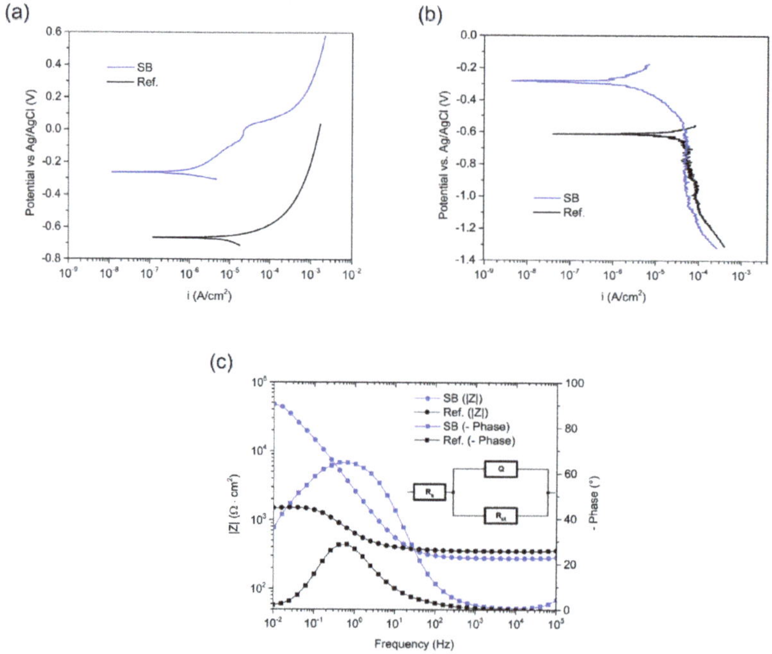

Figure 2. Anodic (**a**) and cathodic (**b**) polarization curves and Bode representation of EIS spectra (**c**) obtained following 3 h of immersion in 0.5 mol/L NaCl solution for bare steel surfaces with and without addition of 3 g/L SB. The inset shows the R(QR) EEC adopted for the spectra modeling.

Table 1. Outcomes of EIS data analysis utilizing the electrochemical equivalence model R(QR).

Sample	R_{ct} ($\Omega \cdot cm^2$)	Y_0 ($S \cdot cm^{-2} \cdot s^n$)	n	R_s ($\Omega \cdot cm^2$)	IE%
Ref.	1.21×10^3	5.45×10^{-4}	8.14×10^{-1}	3.75×10^2	-
SB	6.25×10^4	8.99×10^{-5}	8.08×10^{-1}	2.97×10^2	98

3.2. Prediction of FFC Mitigation Using an Electrochemical Simulated Method

An electrochemical setup to simulate the head and tail of an FFC filament was employed to collect PDPs for the anolyte and catholyte to assess the potential of the synthesized particles to inhibit FFC propagation (Figure 3). In comparison to the polarization curves presented in Figure 2a,b, the anodic and cathodic branches collected in diverse electrolytes could better distinguish the precise contribution of the inhibitor in the specific simulating environment. This approach allowed for the evaluation of the inhibitive effect in terms of the FFC corrosion current ($i_{CORR\ FFC}$), which is determined at the intersection point between the anodic and cathodic branches, as well as the potential difference between the anodic and cathodic sites (ΔE_{CORR}). The output of the measurements is reported in Table 2 for the two different cases under investigation. The curves displayed revealed intriguing outcomes, demonstrating a reduction in ΔE_{CORR} in the order of 87% caused by the presence of SB. The inhibitor's contribution turns out to be overwhelming, acting mainly on the cathodic activity. Besides the shift in the cathodic curve, a decrease in $i_{CORR\ FFC}$ of one order of magnitude is caused by the organic inhibitor's adsorption onto the steel surface. The addition of the inhibitor to the anolyte environment does not yield a significant effect, likely attributable to the aggressive acidic conditions and the lower oxygen content (1.4 ppm [35]). These findings support the characterization of carboxylate inhibitors as non-oxidizing agents, primarily effective in aerated conditions, such as those present in the considered catholyte. Furthermore, the results elucidate how the characterization of the inhibitor's effectiveness, discerned as primarily anodic through electrochemical assessments in nearly neutral solutions and in the literature [13], is contingent upon the specific physicochemical conditions in which it operates. Regarding the presented curves' reproducibility, replicas are provided in Appendix A, Figure A2.

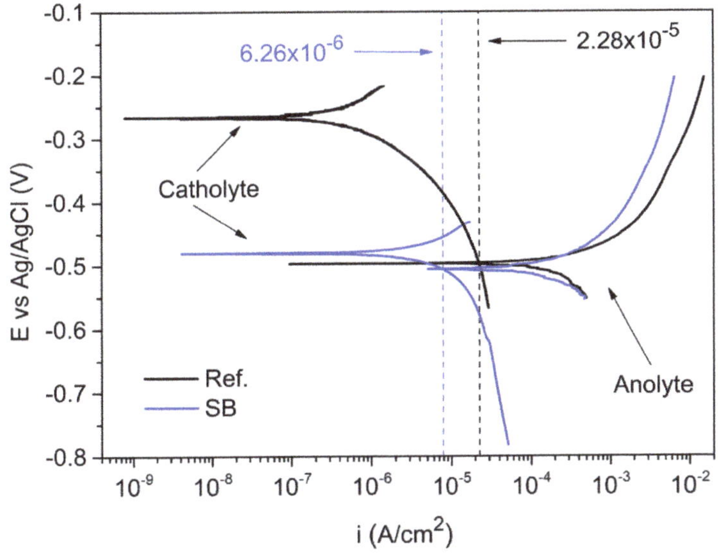

Figure 3. PDPs collected on bare steel surface using anolyte and catholyte "Ref." and with 3 g/L of SB added. The numerical values (expressed in A/cm^2) of FFC corrosion current as the intersection between anodic and cathodic branches are reported.

Table 2. Electrochemical parameters estimated from the electrochemical simulated approach.

Sample	Inhibitor Addition	$i_{CORR\ FFC}$ (A/cm^2)	ΔE_{CORR} (V)
Ref.	-	2.28×10^{-5}	0.230
SB	3 g/L	6.26×10^{-6}	0.030

3.3. *Acrylic-Coated Steel's Durability*

The coated steel samples underwent testing to compare the corrosion evolution of organic layers loaded with SB with those with a neat acrylic coating. FFC tests were conducted under static conditions at 80% relative humidity and 40 °C, revealing a reduction in corrosion propagation from artificial scratches, as depicted in Figure 4. The occurrence of coating delamination, resulting from blister formation and filiform propagation, was observed to be mitigated by the presence of SB within the coating. This is attributed to SB's capacity to migrate through the coating and effectively reach the metallic surface. Notably, discernible disparities in the degradation morphology were observed among the samples, with FFC representing the initial coating detachment event, succeeded by a broader blistering front encompassing the entire initial defect.

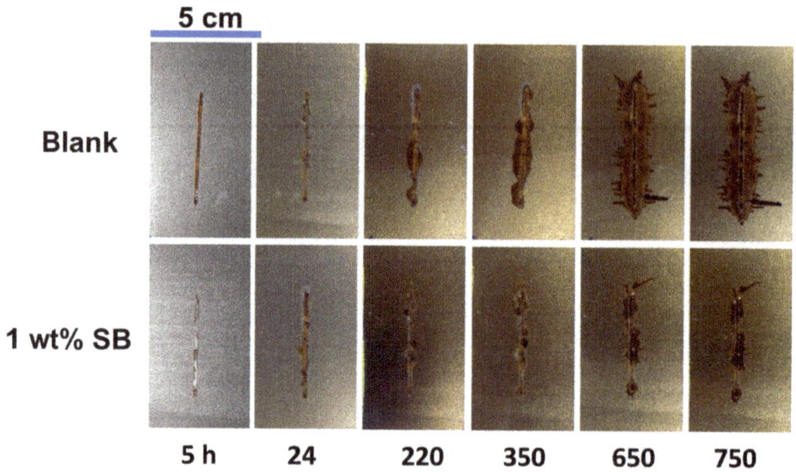

Figure 4. State of corrosion creep of scratched coated samples during 750 h of FFC test at 40 °C in humidostatic conditions (80% r.h.) after 5 h of contamination stage in a neutral salt spray chamber. Acrylic-coated steel panels with (1 wt.% SB) and without (Blank) the addition of the inhibitor are compared.

Figure 5 reports the average value of the expansion of the corrosion-damaged area around the scratch, calculated for three replicates of each type of sample. The inhibition degree was determined to be an average reduction of 35% in the corrosion rate (calculated from the slopes of the trends reported in Figure 5, quantified in terms of the expansion of the corroded area by means of the ImageJ 1.53t software thanks to the transparency of the coating). This result is noteworthy, particularly considering the relatively low concentration of the inhibitor (1 wt.%) incorporated into the coating, in comparison with analogous studies documented in the literature [41–43]. The influence of SB becomes more apparent over time, particularly as filaments begin to emerge due to the initial blister-induced paint delamination. This phenomenon is evident in Figure 4, where the damage observed in two types of samples remains comparable until 220 h of weathering. This observation may be attributed to the reasonable period required for the activation and percolation of the inhibitor through the coating to the interface. Consequently, the incorporation of such organic molecules into the coating does not appear to confer a beneficial effect during the initial stages of FFC. Hence, it might be more advantageous to combine these molecules with other mitigation strategies, such as conversion coatings, which exhibit greater efficacy during the initiation phase [5].

To delve deeper into the behavior of the coating system when exposed to humidity penetration, and also to determine whether the presence of the inhibitor adversely affected the barrier properties of the coating, EIS was conducted on coated steel panels immersed in a 0.5 M NaCl solution. Both monitored samples showed a decrease in values at low frequencies, although the starting values and the impedance decay demonstrate better performance in the case of the coating loaded with SB (Figures 6a and 7a). In the case of the presence of SB in the paint, the transition from capacitive to resistive behavior appears to be less pronounced and sharp compared to the reference "Blank" case and is visible in the spectra starting from 336 h of immersion. This trend could be due to the hydrophobic nature of the interfacial layer formed by the deposits of SB, a property that would hinder the kinetics of water absorption within the organic layer. A transient at the middle frequencies is afforded by the inhibitor, likely due to its solubility and the adsorption process at the metal–paint interface. On the other hand, the "Blank" sample showed a net change between resistive and capacitive behavior.

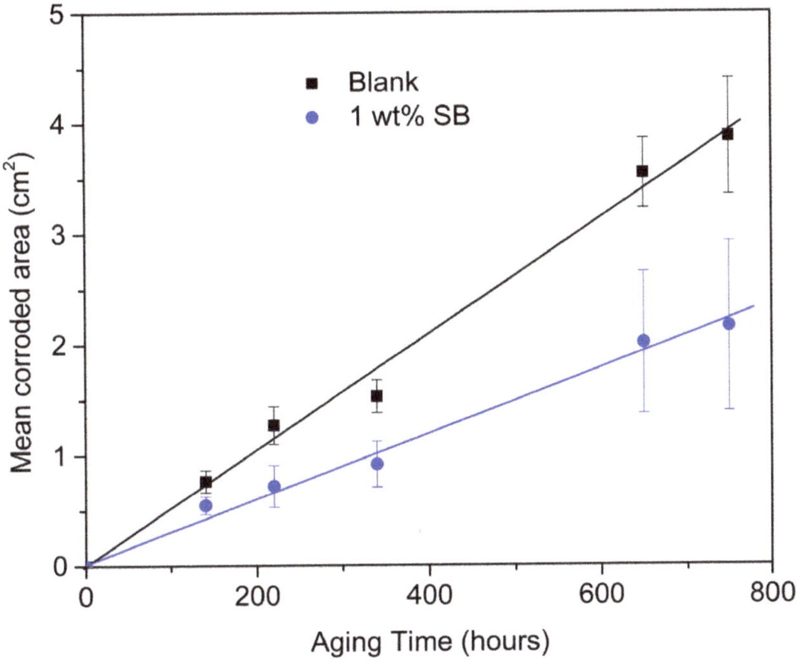

Figure 5. Corroded area evolution over time during 750 h of aging. The mean value calculated for three replicas is reported. Acrylic-coated steel panels with (1 wt.% SB) and without (Blank) the addition of the inhibitor are compared.

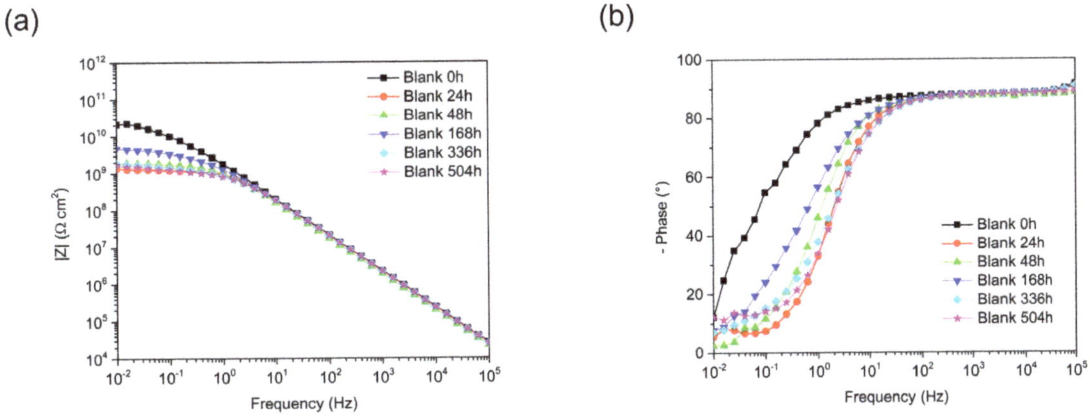

Figure 6. EIS Bode spectra after different immersion periods collected for painted steel samples. Impedance modulus |Z| is reported on the left side (**a**), while a phase angle diagram is shown on the right side (**b**). Blank (neat acrylic paint) behavior is reported during 504 h of immersion in 0.5 M NaCl solution.

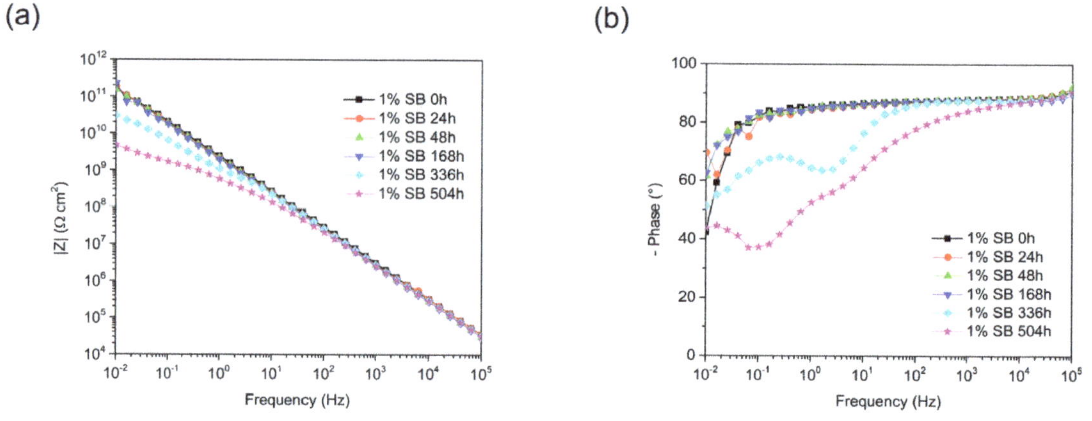

Figure 7. EIS Bode spectra after different immersion periods collected for painted steel samples. Impedance modulus |Z| is reported on the left side (**a**), while a phase angle diagram is shown on the right side (**b**). The behavior of the coating loaded with 1 wt.% SB is reported during 504 h of immersion in 0.5 M NaCl solution.

More in-depth study of the various electrochemical contributions that make up the overall behavior of the coating system was achievable through a spectra modeling study using EECs. However, the results obtained in this step, using a typical organic coating R(Q(R(QR))) data analysis circuit, exhibit in some cases trends that are challenging to attribute to physical phenomena. A CPE (namely Q) is employed to account for the non-ideal capacitive behavior of the coating. Q is helpful in this case for correctly modeling the nonlinearity of the solid electrodes, derived from dissipative phenomena in the polymers, the interaction between polar groups, and diffusive processes at the interface [44]; for this purpose, Q is introduced into the circuit, similar to the case of the EIS collected on bare steel, where its impedance is represented in Equation (1). In principle, two contributions can be attributed to the organic coating (incorporating C_{coat} and R_{coat}) and the interface (composed by the double-layer capacitance C_{dl} and the charge transfer resistance R_{ct}), where the inhibitor adsorbed onto the metal surface should impart a distinguishable effect compared to the reference case, particularly in terms of the charge transfer resistance. This last parameter, as shown in Figure 8a, demonstrates that despite an initial likely higher defectiveness of the "SB" sample, distinguishable from comparison with the reference "Blank" sample, the release of the inhibitor after 150 h of immersion and its subsequent adsorption at the interface provide a healing effect that leads to significantly higher values.

Concerning the other extracted parameters, each interpretation reported in Figure 8 cannot straightforwardly lead to definitive comments on the electrochemical behavior due to the minor differences in values, and therefore any speculation on these parts of the results has been avoided. Only a comparable decrease in the coating resistance R_{coat} for both samples is observed in Figure 8d, indicative of the loss of the barrier properties of the organic layer, together with electrolyte permeation, a feature that does not seem to be altered by the inhibitor's presence in the organic layer.

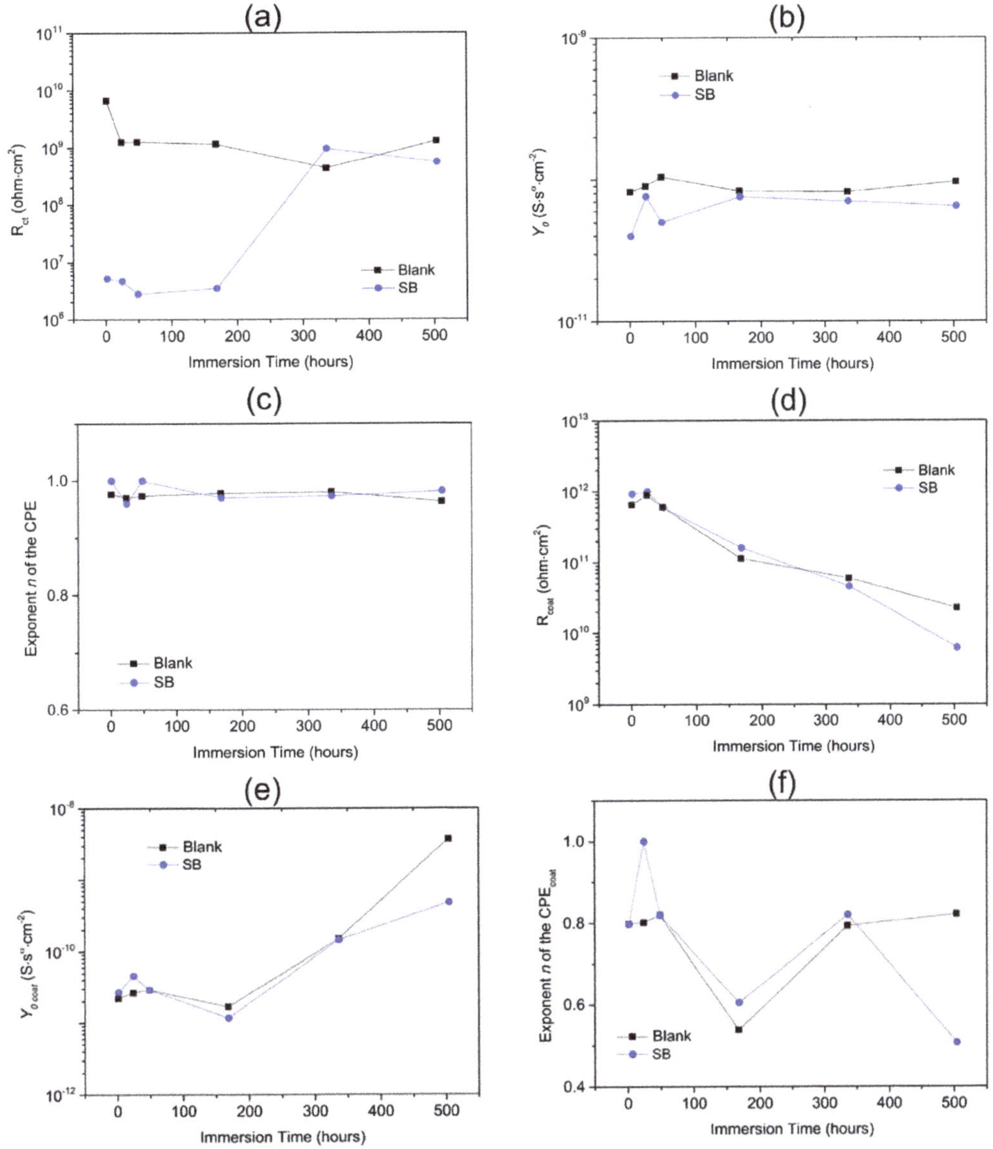

Figure 8. Results of the EIS fitting over testing time. The contributions attributed to the Faradaic processes (**a–c**) and the coating are reported (**d–f**).

4. Discussion

The results unanimously support attributing effective inhibitory power to SB when it is employed on painted steel. It is well known that different corrosion mechanisms lead to varied durability performances for similarly coated substrates, depending on the surrounding environmental conditions, particularly influenced by humidity levels and the presence of contaminant ions. In this study, such evidence is underscored by electrochemical tests based on EIS measurements and PDPs. The diverse chemical composition of the electrolytes used, along with the levels of oxygen present in the solution, resulted in different contributions that the molecule under study made to the evolution of the electrochemical environment at the metal interface. Specifically, SB, known as an anodic inhibitor [13,26], as verified in this study (Figure 2), markedly influences the cathodic activity when an aerated solution exhibits high alkalinity (Figure 3). This scenario occurs in real-life situations where FFC propagates beneath the coating. Consequently, it can be asserted that a singular classification of corrosion inhibitor behavior is not exhaustive, and this characteristic must be related to and studied not only considering the specific case of use, such as organic coatings being exposed in an outdoor environment but also, above all, considering the actual degradation mechanism at play. The choice of an appropriate inhibitory substance should be informed by a detailed understanding of the specific degradation mechanism at play. In the context of preventing coating delamination, it is essential to consider the reactions at the extending front. While some strategies may offer general metal protection irrespective of the corrosion mechanism, others are more tailored to specific phenomena, such as FFC, which is cathodically nucleating and typically exhibits anodic delamination. The selection of a mitigation pathway is therefore contingent on the intricacies of the corrosion process. In the specific case of SB, this substance appeared to be effective in both delamination scenarios, but this is not necessarily a generalizable example and specifically concerns the stability of molecules at different pH levels and their interaction with the metallic substrate in such environments.

5. Conclusions

In this study, we investigated the potential of an eco-friendly organic corrosion inhibitor, namely SB, particularly in the context of acrylic-coated steel. Preliminary electrochemical analyses significantly contributed to our understanding of its inhibitive properties, particularly concerning mild steel in near-neutral conditions. SB is confirmed to act as an anodic corrosion inhibitor under nearly neutral conditions, and electrochemical tests in simulated environments highlight its efficiency in aerated conditions, like other carboxylic acids. In these conditions, an inhibitory effectiveness of 98% was seen compared to situations where the inhibitor was absent.

Additionally, SB has also been shown to limit cathodic activity, specifically in the case of an alkaline, aerated environment, typical of cathodic sites involved in the phenomenon of FFC for coated steel. At this stage, to predict its actual effect on degradation due to FFC, we employed a predictive approach, utilizing electrochemical simulations. This involved analyzing the PDPs of bare steel immersed in test solutions designed to mimic the unique environment anticipated at the metal–paint interface. These simulations yielded results that closely aligned with the observed performance of the coated steel. Specifically, the presence of the inhibitor (referred to as "SB") in the primer layer demonstrated notable limitations in paint delamination and corrosion expansion to 35%.

Author Contributions: Conceptualization, A.C., S.R., F.D. and M.F.; methodology, A.C. and M.F.; software, A.C.; validation, S.R., M.F. and F.D.; formal analysis, A.C., M.F. and S.R.; investigation, A.C.; resources, M.F., S.R. and F.D.; data curation, A.C.; writing—original draft preparation, A.C.; writing—review and editing, A.C., M.F., S.R. and F.D.; visualization, A.C., S.R., F.D. and M.F.; supervision, S.R., F.D. and M.F.; project administration, S.R., F.D. and M.F.; funding acquisition, S.R., F.D. and M.F. All authors have read and agreed to the published version of the manuscript.

Funding: This research received no external funding.

Data Availability Statement: The raw data supporting the conclusions of this article will be made available by the authors upon request.

Acknowledgments: The authors gratefully acknowledge Palini Vernici (Pisogne, BG, Italy), in particular Sergio Bartolomeo.

Conflicts of Interest: The authors declare no conflicts of interest.

Appendix A

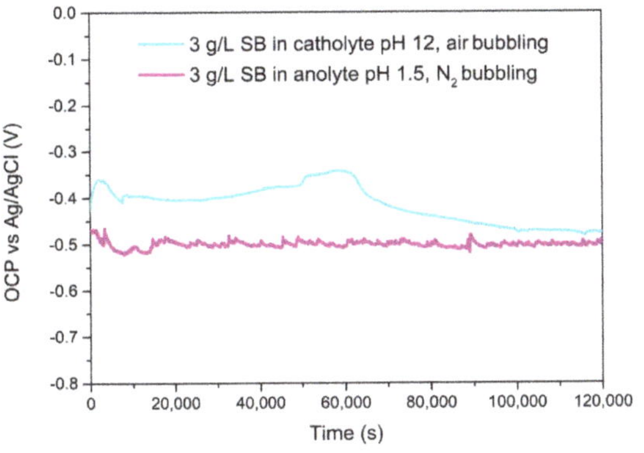

Figure A1. OCP of bare steel immersed in anolyte and catholyte monitored over time with addition of 3 g/L SB.

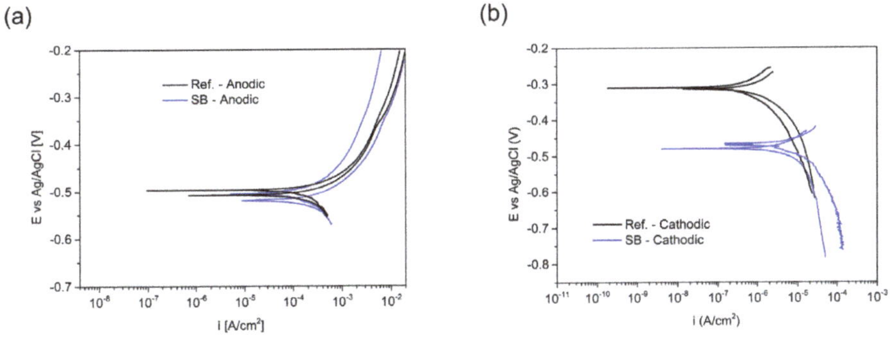

Figure A2. Replicas of the polarization curves obtained in the testing solutions simulating an anodic environment (**a**) and a cathodic one (**b**). PDPs collected on bare steel surface using anolyte and catholyte "Ref." with 3 g/L of SB added.

References

1. Trentin, A.; Pakseresht, A.; Duran, A.; Castro, Y.; Galusek, D. Electrochemical Characterization of Polymeric Coatings for Corrosion Protection: A Review of Advances and Perspectives. *Polymers* **2022**, *14*, 2306. [CrossRef] [PubMed]
2. Lyon, S.B.; Bingham, R.; Mills, D.J. Advances in Corrosion Protection by Organic Coatings: What We Know and What We Would like to Know. *Prog. Org. Coat.* **2017**, *102*, 2–7. [CrossRef]
3. Grundmeier, G.; Schmidt, W.; Stratmann, M. Corrosion Protection by Organic Coatings: Electrochemical Mechanism and Novel Methods of Investigation. *Electrochim. Acta* **2000**, *45*, 2515–2533. [CrossRef]
4. Pélissier, K.; Thierry, D. Powder and High-Solid Coatings as Anticorrosive Solutions for Marine and Offshore Applications? A Review. *Coatings* **2020**, *10*, 916. [CrossRef]
5. Cristoforetti, A.; Rossi, S.; Deflorian, F.; Fedel, M. Exploring the Role of Passivating Conversion Coatings in Enhancing the Durability of Organic-Coated Steel Against Filiform Corrosion Using an Electrochemical Simulated Approach. *Prog. Org. Coat.* **2024**, *189*, 108357. [CrossRef]
6. Milošev, I.; Frankel, G.S. Review—Conversion Coatings Based on Zirconium and/or Titanium. *J. Electrochem. Soc.* **2018**, *165*, C127–C144. [CrossRef]
7. Bastos, A.C.; Ferreira, M.G.; Simões, A.M. Corrosion Inhibition by Chromate and Phosphate Extracts for Iron Substrates Studied by EIS and SVET. *Corros. Sci.* **2006**, *48*, 1500–1512. [CrossRef]
8. Alibakhshi, E.; Ghasemi, E.; Mahdavian, M. Sodium Zinc Phosphate as a Corrosion Inhibitive Pigment. *Prog. Org. Coat.* **2014**, *77*, 1155–1162. [CrossRef]
9. Mahdavian, A.M.; Attar, M.M. Investigation on Zinc Phosphate Effectiveness at Different Pigment Volume Concentrations via Electrochemical Impedance Spectroscopy. *Electrochim. Acta* **2005**, *50*, 4645–4648. [CrossRef]

10. Prosek, T.; Thierry, D. A Model for the Release of Chromate from Organic Coatings. *Prog. Org. Coat.* **2004**, *49*, 209–217. [CrossRef]
11. Kalenda, P.; Kalendová, A.; Štengl, V.; Antoš, P.; Šubrt, J.; Kváča, Z.; Bakardjieva, S. Properties of Surface-Treated Mica in Anticorrosive Coatings. *Prog. Org. Coat.* **2004**, *49*, 137–145. [CrossRef]
12. Saurbier, K.; Schultze, J.W.; Geke, J. Temporary Inhibitors of Corrosion in Wet Atmosphere: Electrochemical Investigations of the Mechanism and Efficiency. *Electrochim. Acta* **1994**, *39*, 1171–1178. [CrossRef]
13. Lahem, D.; Poelman, M.; Atmani, F.; Olivier, M.G. Synergistic Improvement of Inhibitive Activity of Dicarboxylates in Preventing Mild Steel Corrosion in Neutral Aqueous Solution. *Corros. Eng. Sci. Technol.* **2012**, *47*, 463–471. [CrossRef]
14. Rocca, E.; Steinmetz, J. Inhibition of Lead Corrosion with Saturated Linear Aliphatic Chain Monocarboxylates of Sodium. *Corros. Sci.* **2001**, *43*, 891–902. [CrossRef]
15. Georges, C.; Rocca, E.; Steinmetz, P. Synergistic Effect of Tolutriazol and Sodium Carboxylates on Zinc Corrosion in Atmospheric Conditions. *Electrochim. Acta* **2008**, *53*, 4839–4845. [CrossRef]
16. Boisier, G.; Lamure, A.; Pébère, N.; Portail, N.; Villatte, M. Corrosion Protection of AA2024 Sealed Anodic Layers Using the Hydrophobic Properties of Carboxylic Acids. *Surf. Coat. Technol.* **2009**, *203*, 3420–3426. [CrossRef]
17. Shulman, G.P.; Bauman, A.J. Organic Acid Sealants for Anodized Aluminum—A New Method for Corrosion Protection. *Met. Finish.* **1995**, *93*, 16–19. [CrossRef]
18. Hefter, G.T.; North, N.A.; Tan, S.H. Organic Corrosion Inhibitors in Neutral Solutions; Part 1—Inhibition of Steel, Copper, and Aluminum by Straight Chain Carboxylates. *Corrosion* **1997**, *53*, 657–667. [CrossRef]
19. Ormellese, M.; Lazzari, L.; Goidanich, S.; Fumagalli, G.; Brenna, A. A Study of Organic Substances as Inhibitors for Chloride-Induced Corrosion in Concrete. *Corros. Sci.* **2009**, *51*, 2959–2968. [CrossRef]
20. Rammelt, U.; Reinhard, G. Application of Electrochemical Impedance Spectroscopy (EIS) for Characterizing the Corrosion-Protective Performance of Organic Coatings on Metals. *Prog. Org. Coat.* **1992**, *21*, 205–226. [CrossRef]
21. Rammelt, U.; Koehler, S.; Reinhard, G. Electrochemical Characterisation of the Ability of Dicarboxylic Acid Salts to the Corrosion Inhibition of Mild Steel in Aqueous Solutions. *Corros. Sci.* **2011**, *53*, 3515–3520. [CrossRef]
22. Godinez-Alvarez, J.M.; Mora-Mendoza, J.L.; Rodriguez-Betancourt, E.; Zavala- Olivares, G.; Gonzalez-Nunez, M.A. Inhibition of Ferrous Metal Corrosion by Carboxylates. In Proceedings of the CORROSION 2004, New Orleans, Louisiana, 28 March 2004.
23. Mayne, J.; Page, C. Inhibition of the Corrosion of Iron by Benzoate and Acetate Ions. *Br. Corros. J.* **1974**, *9*, 223–226. [CrossRef]
24. Ochoa, N.; Moran, F.; Pébère, N. The Synergistic Effect between Phosphonocarboxylic Acid Salts and Fatty Amines for the Corrosion Protection of a Carbon Steel. *J. Appl. Electrochem.* **2004**, *34*, 487–493. [CrossRef]
25. Caballero, D.; Beltrán-Cobos, R.; Tavares, F.; Cruz-Yusta, M.; Granados, L.S.; Sánchez-Moreno, M.; Pavlovic, I. The Inhibitive Effect of Sebacate-Modified LDH on Concrete Steel Reinforcement Corrosion. *ChemEngineering* **2022**, *6*, 72. [CrossRef]
26. Nguyen, D.T.; To, H.T.X.; Gervasi, J.; Paint, Y.; Gonon, M.; Olivier, M.G. Corrosion Inhibition of Carbon Steel by Hydrotalcites Modified with Different Organic Carboxylic Acids for Organic Coatings. *Prog. Org. Coat.* **2018**, *124*, 256–266. [CrossRef]
27. Bouali, A.C.; Serdechnova, M.; Blawert, C.; Tedim, J.; Ferreira, M.G.S.; Zheludkevich, M.L. Layered Double Hydroxides (LDHs) as Functional Materials for the Corrosion Protection of Aluminum Alloys: A Review. *Appl. Mater. Today* **2020**, *21*, 100857. [CrossRef]
28. Cristoforetti, A.; Deflorian, F.; Rossi, S.; Fedel, M. On the Occurrence of Filiform Corrosion on Organic Coated Carbon Steel Exposed to Cyclic Aging Test. *Corrosion* **2023**, *79*, 1339–1344. [CrossRef] [PubMed]
29. Cristoforetti, A.; Izquierdo, J.; Souto, R.M.; Deflorian, F.; Fedel, M.; Rossi, S. In-Situ Measurement of Electrochemical Activity Related to Filiform Corrosion in Organic Coated Steel by Scanning Vibrating Electrode Technique and Scanning Micropotentiometry. *Corros. Sci.* **2024**, *227*, 111669. [CrossRef]
30. Williams, G.; McMurray, H.N. The Mechanism of Group (I) Chloride Initiated Filiform Corrosion on Iron. *Electrochem. Commun.* **2003**, *5*, 871–877. [CrossRef]
31. Watson, T.M.; Coleman, A.J.; Williams, G.; McMurray, H.N. The Effect of Oxygen Partial Pressure on the Filiform Corrosion of Organic Coated Iron. *Corros. Sci.* **2014**, *89*, 46–58. [CrossRef]
32. Bautista, A. Filiform Corrosion in Polymer-Coated Metals. *Prog. Org. Coat.* **1996**, *28*, 49–58. [CrossRef]
33. Ruggeri, R.T.; Beck, T.R. An Analysis of Mass Transfer in Filiform Corrosion. *Corrosion* **1983**, *39*, 452–465. [CrossRef]
34. Mol, J.M.C.; Hinton, B.R.W.; Van Der Weijde, D.H.; De Wit, J.H.W.; Van Der Zwaag, S. A Filiform Corrosion and Potentiodynamic Polarisation Study of Some Aluminium Alloys. *J. Mater. Sci.* **2000**, *35*, 1629–1639. [CrossRef]
35. Cristoforetti, A.; Rossi, S.; Deflorian, F.; Fedel, M. An Electrochemical Study on the Mechanism of Filiform Corrosion on Acrylic-Coated Carbon Steel. *Prog. Org. Coat.* **2023**, *179*, 107525. [CrossRef]
36. Pourbaix, M.; Burbank, J. Atlas D-Equilibres Electrochimiques. *J. Electrochem. Soc.* **1964**, *111*, 14C. [CrossRef]
37. *ASTM D2803-09*; Standard Guide for Testing Filiform Corrosion Resistance of Organic Coatings on Metal. ASTM International: West Conshohocken, PA, USA, 2020.
38. *ASTM B117-19*; Standard Practice for Operating Salt Spray (Fog) Apparatus. ASTM International: West Conshohocken, PA, USA, 2019.
39. Cristoforetti, A.; Parola, F.; Parrino, F.; Izquierdo, J.; Souto, R.M.; Rossi, S.; Deflorian, F.; Fedel, M. Sebacate Intercalated CaAl Layered Double Hydroxide Pigments for Corrosion Protection of Low Carbon Steel: Anion Exchange and Electrochemical Properties. *Appl. Clay Sci.* **2024**, *250*, 107300. [CrossRef]
40. Fedel, M.; Poelman, M.; Olivier, M.; Deflorian, F. Sebacic Acid as Corrosion Inhibitor for Hot-Dip Galvanized (HDG) Steel in 0.1 M NaCl. *Surf. Interface Anal.* **2019**, *51*, 541–551. [CrossRef]

41. Peng, Y.; Hughes, A.E.; Mardel, J.I.; Deacon, G.B.; Junk, P.C.; Catubig, R.A.; Forsyth, M.; Hinton, B.R.W.; Somers, A.E. Dual Function of Rare Earth Carboxylate Compounds on the Barrier Properties and Active Corrosion Inhibition of Epoxy Coatings on Mild Steel. *Prog. Org. Coat.* **2023**, *185*, 107870. [CrossRef]
42. Catubig, R.; Seter, M.; Neil, W.; Forsyth, M.; Hinton, B. Effects of Corrosion Inhibiting Pigment Lanthanum 4-Hydroxy Cinnamate on the Filiform Corrosion of Coated Steel. *J. Electrochem. Soc.* **2011**, *158*, C353–C358. [CrossRef]
43. Glover, C.F.; Williams, G. Inhibition of Corrosion-Driven Organic Coating Delamination and Filiform Corrosion on Iron by Phenyl Phosphonic Acid. *Prog. Org. Coat.* **2017**, *102*, 44–52. [CrossRef]
44. Tokutake, K.; Nishi, H.; Ito, D.; Okazaki, S.; Serizawa, Y. Relationship between Degradation Characteristics of Organic Coating on Internal Bottom Plate of Oil Storage Tank and Constant-Phase Element Parameter Values. *Prog. Org. Coat.* **2015**, *87*, 69–74. [CrossRef]

metals MDPI

Article

Low-Carbon Steel Formed by DRECE Method with Hot-Dip Zinc Galvanizing and Potentiodynamic Polarization Tests to Study Its Corrosion Behavior

Jiřina Vontorová [1,*], Vlastimil Novák [1] and Petra Váňová [2]

[1] Department of Chemistry and Physico-Chemical Processes, Faculty of Materials Science and Technology, VSB-Technical University of Ostrava, 708 00 Ostrava, Czech Republic; vlastimil.novak@vsb.cz

[2] Department of Materials Engineering and Recycling, Faculty of Materials Science and Technology, VSB-Technical University of Ostrava, 708 00 Ostrava, Czech Republic; petra.vanova@vsb.cz

* Correspondence: jirina.vontorova@vsb.cz

Abstract: The use of low-carbon unalloyed steel with minimal silicon content is widespread in structural steel and automotive applications due to its ease of manipulation. The mechanical properties of this steel can be significantly enhanced through severe plastic deformation (SPD) techniques. Our study focuses on the practical benefits of the dual rolling equal channel extrusion (DRECE) method, which strengthens the steel and has implications for material hardness and the thickness of subsequently applied hot-dip zinc galvanizing. Furthermore, the steel's corrosion potential and current are investigated as a function of material hardness and thickness. The findings show a 20% increase in hardness HV 30 after the first run through the forming machine, with an additional 10% increase after the second run. Subsequent galvanizing leads to a further 1–12% increase in HV 30 value. Notably, the DRECE hardening demonstrates no statistically significant effect on the corrosion potential and current; however, the impact of galvanizing is as anticipated.

Keywords: severe plastic deformation; DRECE method; hot dip galvanizing; Zn coating; GDOES; optical microscopy; corrosion; Tafel plot

1. Introduction

DC03 steel is a traditional steel that is appreciated for its ductility characteristics. With its low carbon content, it offers increased elasticity and is easier to shape, bend, weld, and cut. This makes it an ideal option for creating structural components and intricate forms. DC03 steel is commonly utilized in the production of metal furniture, automotive body parts, a variety of industrial components, and electronic equipment.

The DRECE (dual rolling equal channel extrusion) method is part of the SPD (severe plastic deformation) process utilized to enhance the mechanical properties of metals and produce ultrafine-grained (UFG) materials. Through the SPD process, grain sizes of 20 to 200 nm can be achieved [1]. The resulting grain size can be influenced by adjusting the forming parameters (for example, by increasing the deformation pressure or reducing the material temperature during forming) [2]. Additionally, the chemical composition of the material formed significantly influences the resulting grain size [3].

The DRECE forming process involves intensive plastic deformation of the material. The prototype device, located at the Department of Mechanical Technology, Faculty of Mechanical Engineering, VŠB-Technical University of Ostrava, is based on the DCAP method [4–6] (see Figure 1). Figure 1b shows the deformation zone rotated by 180° in the "z" axis (vertical direction) compared to Figure 1a. In 2024, a patent was granted for a new design of a shaping device intended for industrial use.

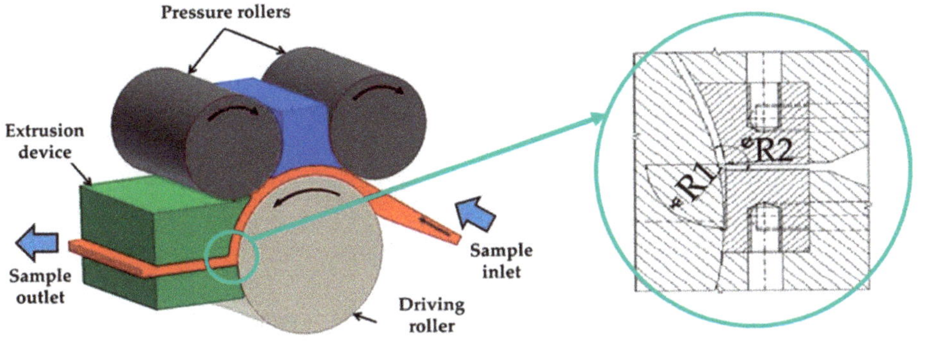

a b

Figure 1. Schematics of the DRECE forming method (adapted from [7]): (**a**) schematics of the equipment and extrusion principle; and (**b**) schematics of the deformation zone.

The main parts of the device are two pressure cylinders, which allow the regulation of the pressure force, and one feed (drive) cylinder. The cylinders are driven by an electric motor with a NORD gearbox via a vane clutch. The belt feed speed range is v = 1–12 mm/s, and the pressure of the rollers can be adjusted in the following range: front roller: p1 = 15–20 MPa, rear roller: p2 = 5–10 MPa, and the motor power input is 0.236 kW.

A piece of sheet metal is fed into the processing area and is pushed through the forming tool by the feed roller, working in tandem with the pressure rollers, without altering the cross-section. The structure can be significantly refined by subjecting the material to multiple plastic deformations in this manner [8]. In reference [7], the authors analyze the impact of severe plastic deformation (SPD) on the quality of DC03 steel at forming tool angles $\Phi = 113^\circ$. The angle at which the forming tool is set in the deformation zone is the most critical factor in determining the efficiency of the forming process, as it enables high deformation intensity and the reshaping of the material [6].

Additional important considerations include the radii R1 and R2 (Figure 1b). Their values are influenced by the characteristics of the strip material. The objective is to achieve shear deformation for grain refinement while ensuring that stress concentrations do not exceed the material's tolerance. In terms of stress distribution, the magnitude of the shear stress must surpass the magnitude of the bending stress during the application of the DRECE method.

In reference [9], the authors describe the effect of increasing the feed rate of the steel strip from 10 mm/s to 100 mm/s, which significantly increased the efficiency of the forming process. Structural analysis on TEM demonstrated a reduction in the mean grain size of the steel strip.

In [10], the authors describe the modernization of the pressing equipment. The deformation zone angle was reduced from 113° to 108°. The mechanical properties after the first pass through the welding machine were increased by 60% for the yield strength Re and by 40% for the ultimate strength Rm. Similar values were also achieved for steel strip DC01 [9].

Hot-dip zinc galvanizing is widely recognized as a highly effective method for safeguarding steel against corrosion [11,12]. This technique facilitates the creation of premium coatings, thereby offering long-term corrosion protection [13] at a reasonable operating cost [14,15]. These coatings serve as a barrier on the steel surface, providing protection against corrosion. Steel corrosion refers to the degradation of metal caused by electrochemical or chemical reactions in a corrosive environment [16].

The process of hot-dip galvanizing steel involves creating an alloy coating of hot-dip zinc on the steel surface. This process involves a series of metallurgical reactions, diffusion, and thermodynamic changes. Standardization for the hot-dip galvanizing of steel parts can be found in EN ISO 1461 [17] and EN ISO 14713-2 [18]. The quality of the galvanized surface is primarily influenced by the chemical composition. The low silicon content plays a crucial role [19,20], as illustrated by the Sandelin diagram (refer to Figure 2). With regard to the impact of the silicon content on the hot-dip galvanizing process, steels can be categorized based on their silicon content levels as follows [21]: steels with low silicon content (<0.03 wt.% Si); steels falling within the so-called "Sandelin region" with a silicon content in the range of 0.03–0.12 wt.%; steels in the "Sebisty region" with silicon content around (~0.15–0.25 wt.%); and steels with a high silicon content (>0.28 wt.%). Steels with a high silicon content are not recommended for use in the hot-dip galvanizing process due to the resulting thick and brittle coatings, low mechanical resistance, and susceptibility to delamination.

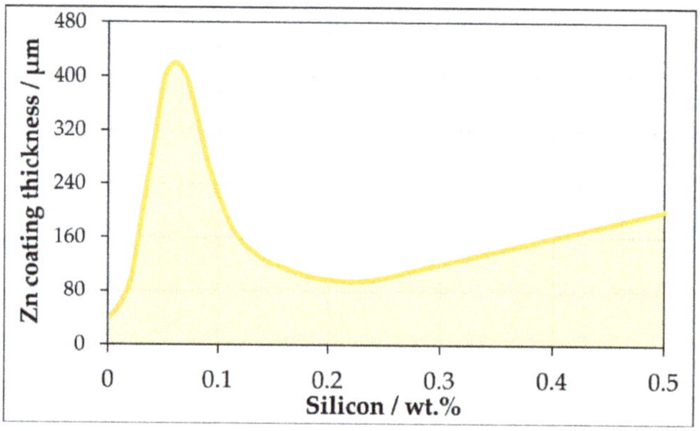

Figure 2. Sandelin diagram.

Depending on the chemical composition of the steel, the temperature of the galvanizing bath, the composition of the bath, the dwell time in the bath, the surface quality and wall thickness of the galvanized material, and the method and rate of cooling, a coating composed of different Fe-Zn intermetallic compounds is formed [22]. These compounds form layers with different grain compositions, morphologies, thicknesses, and mechanical properties [23]. The uppermost layer contains nearly pure zinc, referred to as the η-phase [24]. Beneath this are the monoclinic ζ-phase crystals [25–27], followed by a consistent layer of closely arranged hexagonal δ-phase crystals [27–30]. The bottommost is a thin layer of Γ-phase [31–33].

In the literature [9,10], there is a description of the impact of DRECE forming on the enhancement of steel hardness. However, there is a lack of information regarding the hot-dip galvanizing surface treatment of such a hardened material. This study seeks to address this gap by examining the properties of hardened steel subjected to hot-dip galvanizing.

2. Materials and Methods

For this research, DC03, a low-carbon, low-silicon steel, was selected after applying the DRECE method. The standard for identifying DC03 steel is 1.0347. The low silicon content is crucial for enabling hot-dip galvanizing. The chemical composition of this steel, determined by the BULK GDOES method using a Spectruma Analytik GMBH (Hof, Germany) glow discharge optical emission spectrometer (GDOES) under 700 V and 35 mA excitation conditions [34,35], is presented in Table 1. The elements marked with an asterisk are determined through elemental analysis using an Eltra (Haan, Germany) CS 2000 combustion analyzer [36,37].

Table 1. Chemical composition of DC03 steel.

C*	Mn	Si	P	S*	Cr	Ni	Mo	Cu
				wt%				
0.0406	0.170	0.007	0.013	0.0140	0.028	0.037	0.002	0.061
Ti	**Co**	**B**	**Pb**	**V**	**W**	**Al**	**Nb**	
				wt%				
<0.001	0.006	0.0004	<0.001	0.001	<0.001	0.034	<0.001	

The specimens, which consist of strips of sheet metal measuring 58 mm in width, 2 mm in thickness, and 2000 mm in length, undergo extrusion using the DRECE experimental apparatus (refer to Figure 3). The process parameters are configured at the lower limits of the ranges as mentioned above.

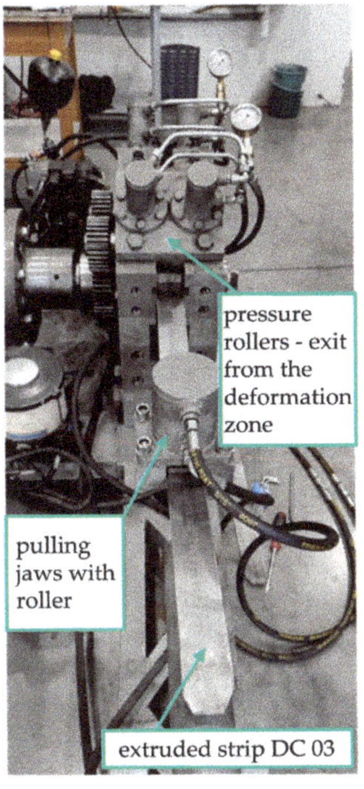

Figure 3. Shaping device. Test progress (photo by author).

The specimens, each approximately 1 m in length, were then subjected to a standard galvanizing process at a hot-dip galvanizing plant. This involves a series of steps, including degreasing, rinsing, pickling with dilute hydrochloric acid, rinsing, immersion in flux, drying, and, finally, immersion in molten zinc, as shown in Figure 4.

Table 2 lists the samples with DRECE and surface treatment. In addition, Figure 5 displays a photograph of the surfaces of all the samples. It is worth noting that the samples without the surface treatment exhibit grooves left by the forming tool, which could potentially impact the mechanical properties of the material being evaluated. Conversely, the galvanizing process results in the formation of a uniform, compact, smooth layer.

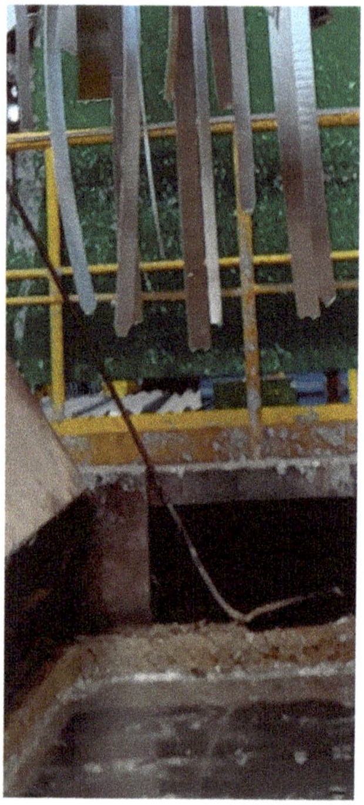

Figure 4. Samples after hot-dip galvanizing (photo by author).

Table 2. Description of samples with DRECE and surface treatment.

Without Surface Treatment		Hot-Dip Galvanized	
0	DC03	Z0	DC03
1	DC03, 1× DRECE	Z1	DC03, 1× DRECE
2	DC03, 2× DRECE	Z2	DC03, 2× DRECE
2R	DC03, 2× DRECE with rotation	Z2R	DC03, 2× DRECE with rotation

In the provided description, it can be seen that the microscopic steel images (Figure 6) of samples 0 and 2, denoted as samples in their original state (marked 0) and after two passes through the shaping device (marked 2), with and without the surface treatment (marked Z0 and Z2), correspond to specific structural characteristics. The samples were analyzed in longitudinal and transverse sections, embedded in resin, ground, and polished. The etching was conducted with a Nital 4% reagent. Microstructure observation and zinc layer thickness measurements (see Section 3.1) were performed with an Olympus (Tokyo, Japan) IX70 optical microscope at 200× magnification. The results were analyzed using Image Pro Micro G software (ver. 10).

Upon examination, it was determined that the microstructure of the initial state sample (sample 0) was ferritic, featuring an average grain size of 30 μm, corresponding to a grain size number G = 7 according to standard ISO 643:2021-04 [38]. At this stage, the grains exhibit distinct boundaries. Following zinc galvanizing, the disappearance of certain grain boundaries was noted (sample Z0). A similar microstructure was also observed in sample 2 subsequent to the double passage by the DRECE method. Furthermore, after the zinc galvanizing of the reshaped part, the etchability of the grain boundaries was once again reduced (sample Z2). Notably, the etchability of the microstructure appeared to deteriorate due to the increased dislocation density and the corrosion protection effect of the zinc coating.

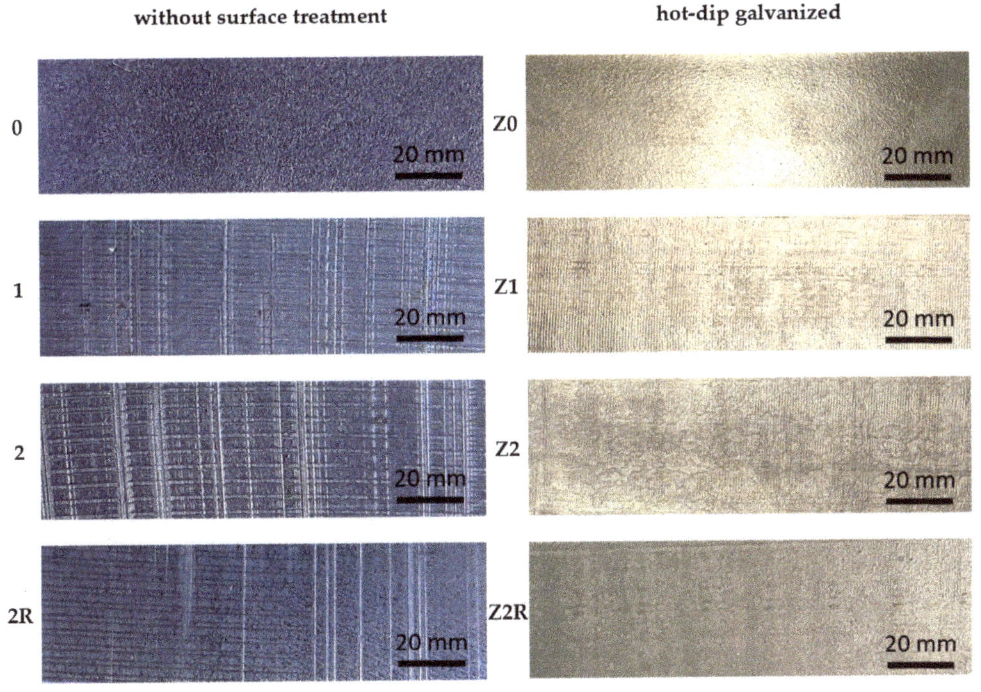

Figure 5. Images of samples under DRECE and surface treatment.

Hardness measurements were performed using the Vickers hardness test according to standard ISO 6507-1:2023 [39]. The measurements were carried out using an EMCO-Test (Kuchl, Austria) DuraScan G5 micro-hardness tester.

The polarization analysis of the steel samples was conducted using the Tafel extrapolation method with a Metrohm Autolab (Utrecht, The Netherlands) PGSTAT302 potentiostat in the standard three-electrode configuration. In this setup, the steel sample functions as the working electrode, an argent chloride (Ag/AgCl) electrode acquired from 2THETA (CR) serves as the reference electrode, and a platinum wire serves as the counter electrode. All experiments were conducted at ambient temperature (25 °C) in 0.1 M HCl, 0.1 M H_2SO_4, and 5% NaCl electrolytes, with chemicals procured from MACH CHEMICALS Ltd. (Ostrava, Czechia); the solutions were diluted using deionized water. The steel samples were immersed in the electrolyte to a depth of 1 cm, and precise adjustment of the immersion was achieved using a micrometer screw. The solutions were purged by nitrogen gas (Messer, Ostrava, Czechia) for 30 min prior to each electrochemical test. The potential was measured at a rate of 2 $mV{\cdot}s^{-1}$ over a potential range of −1.5 V to 1.0 V. At low polarization rates, the measured currents and voltages convey information about the corrosion process occurring at the interface of the material and electrolyte. In contrast, at high polarization rates, the measured current not only reflects the current value during the corrosion process, but also encompasses the charge of the surface capacitor. Therefore, if the scan rate is not sufficiently low, the current density will exceed that which arises from the corrosion reactions [40]. Although a potential scan rate of 2 $mV{\cdot}s^{-1}$ was applied, no distortions into the potentiodynamic polarization curves were observed, as has been previously reported [41,42]. No distortions to determine the corrosion current density were observed.

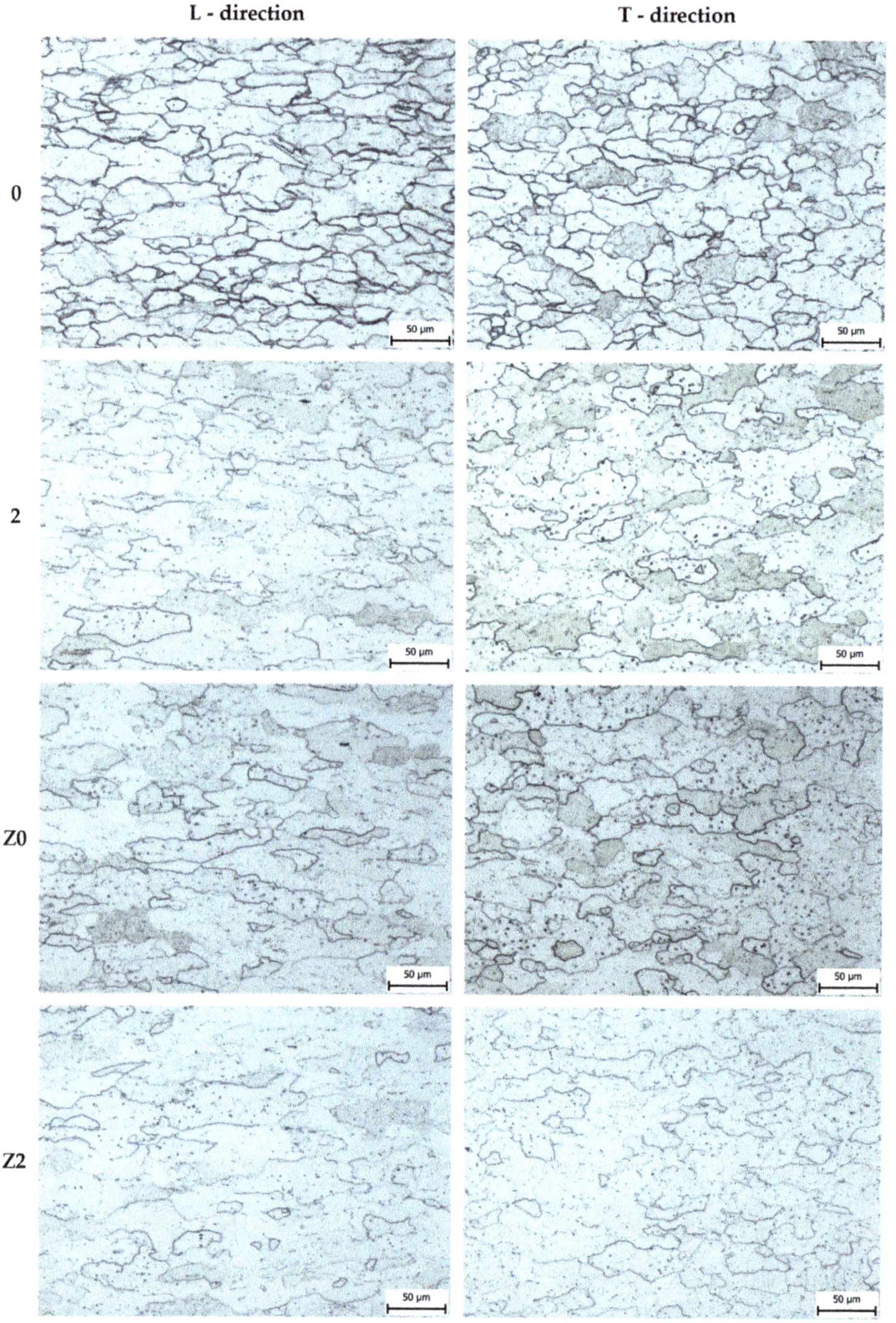

Figure 6. Microscopic images of samples in the longitudinal direction (**L**) and in the transverse direction (**T**).

3. Results and Discussion

3.1. Thickness of the Zinc Layer

The accompanying images display the zinc coating on the base material without DRECE hardening. In Figure 7a, the macroscopic view of the zinc coating on the steel specimen is presented in both the longitudinal (L) and transverse (T) directions. The images depict an impressively even distribution of the zinc coating. Subsequently, Figure 7b illustrates measurements of the zinc coating thickness on samples prepared in both directions.

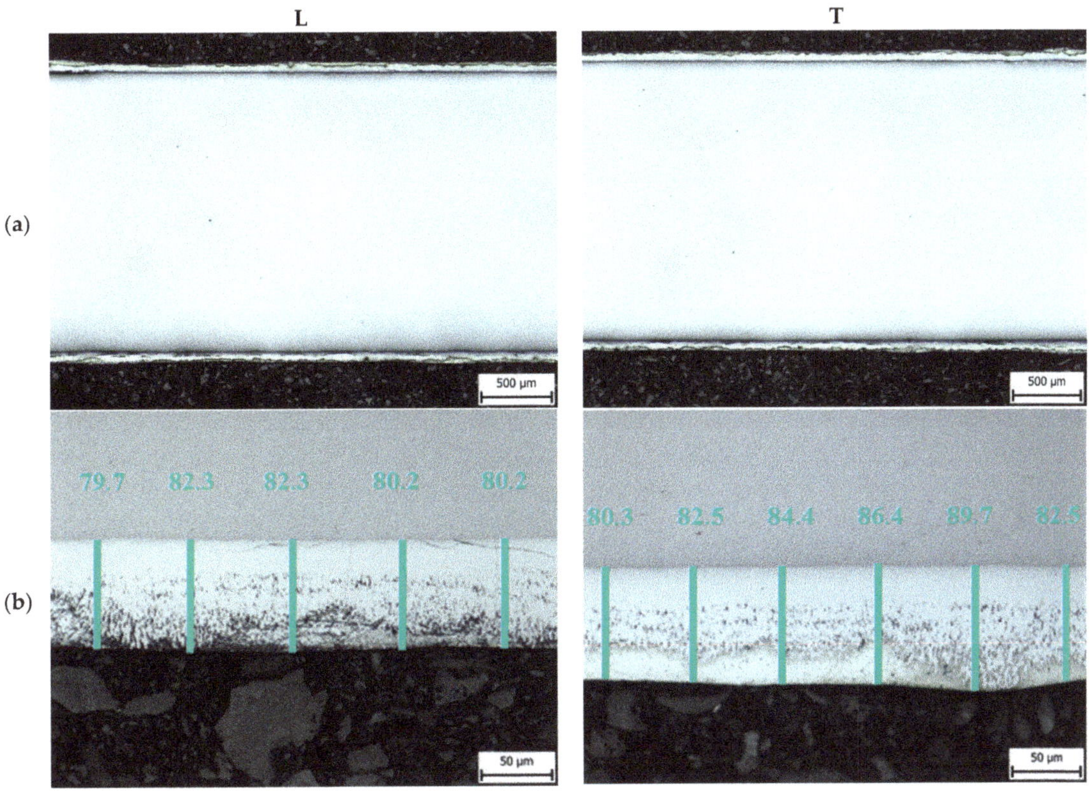

Figure 7. Zn coating on sample Z0 (**a**); and measurement of Zn coating thickness on sample Z0 (**b**). Optical microscopy.

A suitable alternative to optical microscopy for determining the thickness of zinc coatings on planar samples with smooth surfaces is the profile GDOES (Glow Discharge Optical Emission Spectroscopy) analysis. The analysis was performed under the following excitation conditions: 1000 V and 15 mA [43] directly from the surface of the Z0 sample (galvanized sample without DRECE hardening (Figure 8)), 20 mm from the edge of the sample, without sample preparation (cutting, resin potting, polishing, etching). Due to surface heterogeneities, performing GDOES analysis of the galvanized specimens is impossible after DRECE forming.

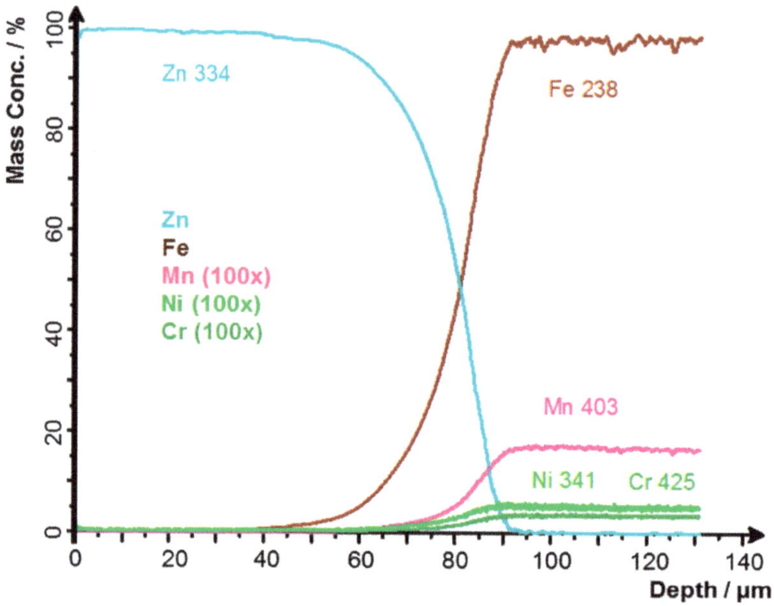

Figure 8. Zn coating thickness measurement on sample Z0, profile GDOES analysis.

The profile curves of zinc and other elements agreed with the results of optical microscopy, revealing a zinc coating thickness of around 80 µm. The measured concentrations also matched those obtained from BULK GDOES analysis (Table 1). It is worth noting that the values extracted from the record must be divided by the number indicated in brackets in the legend of the profile record. The concentrations of certain elements are so low that if the *y*-axis scale remained unchanged, they would be undetectable and indistinguishable from one another.

Figure 9 displays the Zn coating on the steel after DRECE hardening. Additionally, Table 3 illustrates the average layer thicknesses. According to the findings, DRECE hardening results in a minor reduction in the thickness of the subsequent hot-dipped Zn coating. Nevertheless, this reduction is not considered statistically significant.

Table 3. Average thicknesses of Zn coatings/µm.

Sample Label	Thickness of Zn Coating/µm	
	L	T
Z0	80.9 ± 3.1	84.3 ± 3.5
Z1	78.9 ± 3.9	83.6 ± 5.2
Z2	72.5 ± 3.8	71.2 ± 5.3
Z2R	70.7 ± 3.5	75.4 ± 3.8

The coating microstructure corresponds to the requirements for the quality of the hot-dip galvanizing surface treatment. All intermetallic phases, i.e., η, ζ, δ and Γ phases, were identified in the microstructure of the zinc coating. Figure 9b shows the delamination of the zinc coating.

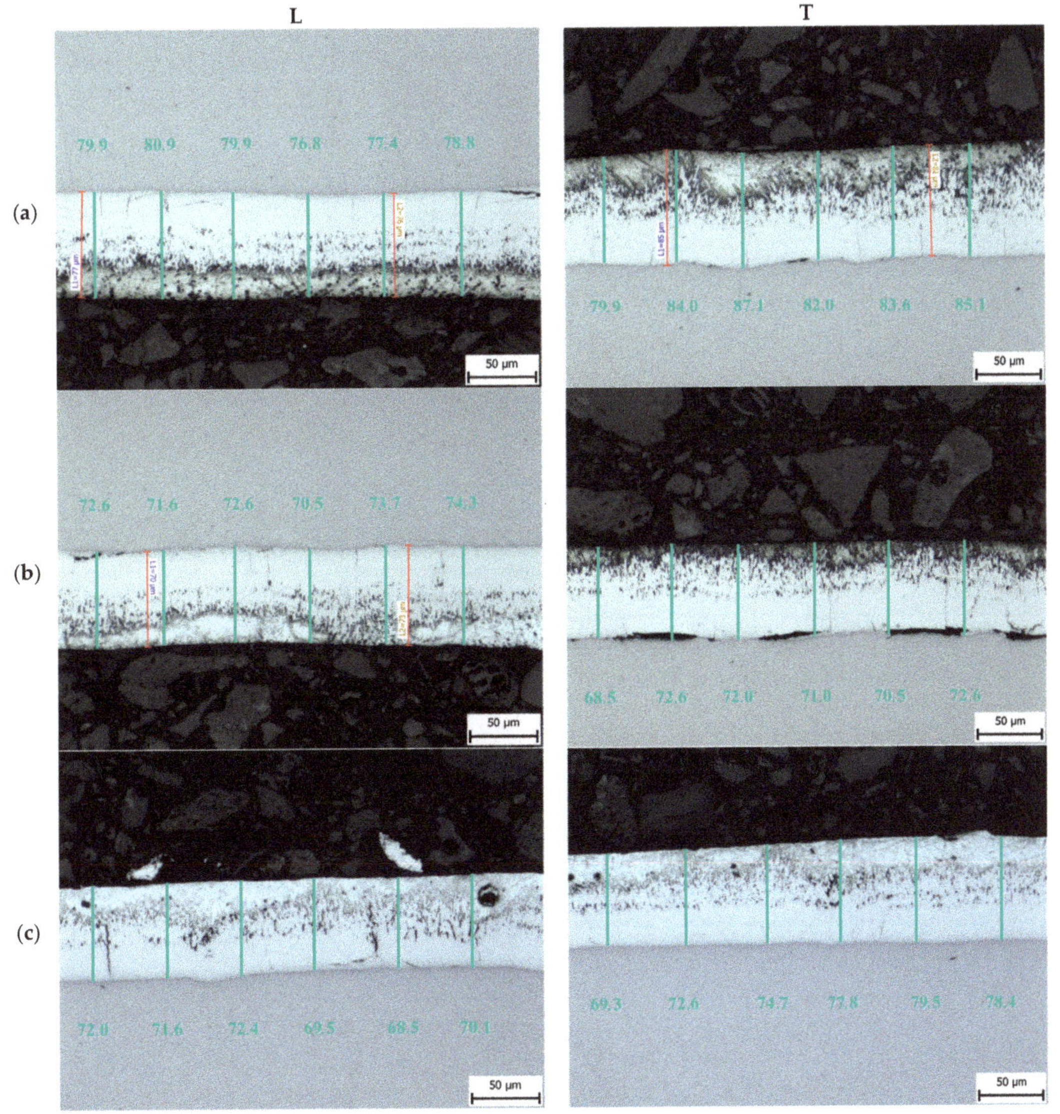

Figure 9. Measurement of Zn coating thickness on sample Z1 (**a**); Z2 (**b**); and Z2R (**c**). Optical microscopy, magnification 200×.

3.2. HV 5 Hardness

The HV 5 hardness of all examined samples was assessed to evaluate the SPD's efficiency. Hardness measurements were made on metallographic sections at the center of the plate thickness. As depicted in Figure 10, the chart illustrates an 18% increase in the base material's hardness after a single pass through the forming machine, followed by an additional 12% increase after a second pass. Subsequent galvanization resulted in a further augmentation of HV 5 by 1–10%.

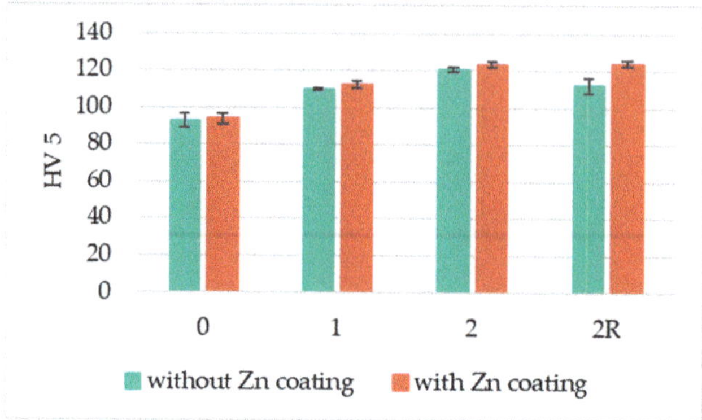

Figure 10. HV 5 hardness of all samples in the L direction.

3.3. Electrochemical Testing

For the electrochemical testing of corrosion behavior, the Tafel extrapolation method is a suitable technique and provides significant time savings compared to the classical immersion method based on mass loss determination. This technique allows for the extrapolation of information relating to corrosion current density (i_{corr}), corrosion potential (E_{corr}), anodic (β_a) and the cathodic (β_c) polarization curve slope and corrosion rate. The measurements are conducted using the linear sweep voltammetry technique, recording current density at varying applied potentials. Subsequently, the recorded data are converted to a logarithm of the current density versus voltage, facilitating the extrapolation process.

In the provided study, Figures 11–13 exhibit the Tafel plots illustrating the behavior of the prepared steel samples in various electrolytes, specifically, 0.1 M HCl, 0.1 M H_2SO_4, and 5% NaCl. In addition, Figure 14 shows repeated measurements of sample 0 in 0.1 M HCl, including Tafel extrapolation. The corrosion potential values for the untreated and DRECE-treated steel samples are relatively consistent, averaging around −0.40 V and −0.41 V when exposed to the corrosive influence of hydrochloric acid and sulfuric acid, respectively. However, a noteworthy shift to −1.07 V was noted for the samples exposed to 5% NaCl. Furthermore, samples subjected to hot-dip galvanizing displayed a noticeable leftward shift in the corrosion potentials compared to the ungalvanized samples, indicating the formation of a protective zinc coating.

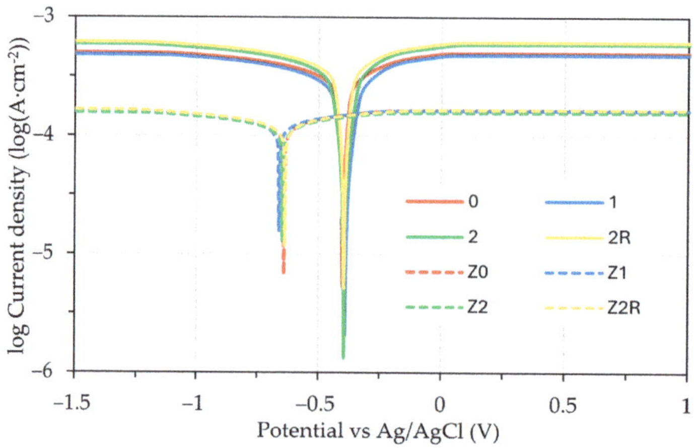

Figure 11. Tafel curves of the tested steel samples in 0.1 M HCl solution.

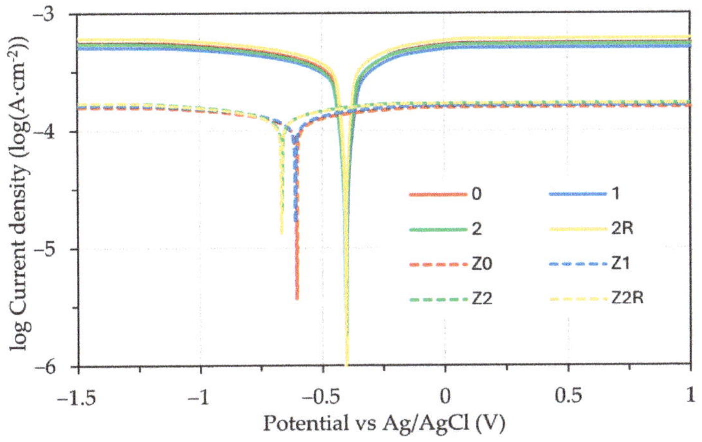

Figure 12. Tafel curves of the tested steel samples in 0.1 M H_2SO_4 solution.

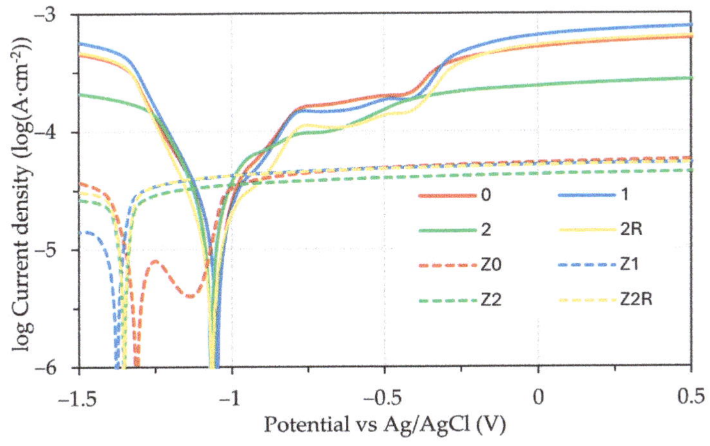

Figure 13. Tafel curves of the tested steel samples in 5% NaCl solution.

Additionally, significant variations in the corrosion current and corrosion rate are observed based on the treatment method and the specific electrolytic medium used. Notably, galvanized steels exhibit markedly lower values for these parameters, underscoring the enhanced corrosion resistance associated with this treatment method.

The data presented in Table 4 illustrate the Tafel extrapolation results for the steel samples tested in three different media. The findings indicate a significant enhancement of corrosion resistance for the hot-dip galvanized steels, as evidenced by the computed corrosion rates. It is noteworthy that the slopes of the anodic polarization curves (β_a) are generally steeper than those of the cathodic polarization curves (β_c), suggesting that the electrochemical polarization of the samples is predominantly governed by anodic processes, with the anodic reaction being more dominant than the cathodic reaction. However, it is important to note that exceptions were observed for samples tested in 5% NaCl. In addition, the polarization resistance values were significantly higher in the samples that underwent hot-dip galvanizing, attributed to the development of a protective layer on the surface. However, it is worth noting that exceptions were observed for samples tested in 5% NaCl.

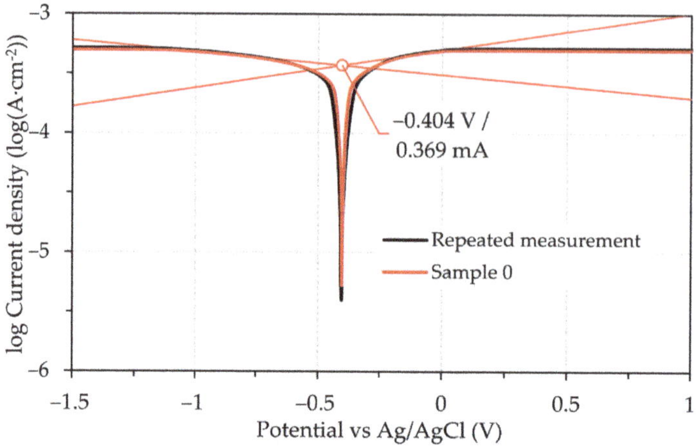

Figure 14. Repeated recording of the polarization potentiodynamic curve of sample 0 in 0.1 M HCl. The original recording shows the Tafel extrapolation.

Table 4. Parameters of Tafel extrapolation for tested samples in different electrolytes.

	0	**1**	**2**	**2R**	**Z0**	**Z1**	**Z2**	**Z2R**
				0.1 M HCl				
E_{corr} (V)	−0.404 ± 0.0471	−0.394 ± 0.0305	−0.404 ± 0.0316	−0.404 ± 0.0522	−0.641 ± 0.0312	−0.663 ± 0.0311	−0.649 ± 0.0304	−0.642 ± 0.0369
i_{corr} (mA·cm^{-2})	0.369 ± 0.0300	0.350 ± 0.0315	0.422 ± 0.0311	0.447 ± 0.0314	0.127 ± 0.0374	0.125 ± 0.0309	0.122 ± 0.0301	0.128 ± 0.0319
β_a (V·decade^{-1})	0.315	0.333	0.326	0.315	0.291	0.310	0.277	0.259
β_c (V·decade^{-1})	−0.192	−0.197	−0.196	−0.190	−0.225	−0.248	−0.219	−0.215
Corrosion rate (mm·year^{-1})	4.376	4.143	4.999	5.300	1.504	1.478	1.449	1.515
Polarization resistance (Ω)	2318	2348	1972	1922	6638	6240	7154	7153
				0.1 M H_2SO_4				
E_{corr} (V)	−0.409 ± 0.0570	−0.405 ± 0.0500	−0.411 ± 0.0420	−0.409 ± 0.0549	−0.610 ± 0.0403	−0.614 ± 0.0523	−0.675 ± 0.0439	−0.672 ± 0.0452
i_{corr} (mA·cm^{-2})	0.403 ± 0.0451	0.374 ± 0.0410	0.394 ± 0.0437	0.441 ± 0.0400	0.117 ± 0.0461	0.121 ± 0.0427	0.132 ± 0.0517	0.130 ± 0.0423
β_a (V·decade^{-1})	0.316	0.303	0.321	0.308	0.368	0.384	0.296	0.302
β_c (V·decade^{-1})	−0.199	−0.188	−0.193	−0.187	−0.323	−0.317	−0.248	−0.244
Corrosion rate (mm·year^{-1})	4.774	4.437	4.673	5.225	1.384	1.428	1.559	1.539
Polarization resistance (Ω)	2093	2360	2141	1991	5378	5141	6063	6132
				5% NaCl				
E_{corr} (V)	−1.052 ± 0.0327	−1.049 ± 0.0310	−1.070 ± 0.0596	−1.062 ± 0.0321	−1.304 ± 0.0524	−1.364 ± 0.0375	−1.366 ± 0.0300	−1.364 ± 0.0356
i_{corr} (mA·cm^{-2})	0.018 ± 0.0331	0.017 ± 0.0302	0.019 ± 0.0422	0.011 ± 0.0343	0.004 ± 0.0380	0.004 ± 0.0302	0.004 ± 0.0304	0.004 ± 0.0409
β_a (V·decade^{-1})	3.764	3.674	4.119	4.363	7.535	31.218	14.500	11.622
β_c (V·decade^{-1})	−4.237	−4.862	−4.894	−5.549	−8.728	−6.952	−24.901	−22.506
Corrosion rate (mm·year^{-1})	0.211	0.201	0.230	0.125	0.048	0.049	0.046	0.049
Polarization resistance (Ω)	3044	3004	2488	4170	6602	2737	2850	3085

4. Conclusions

This research delves into the impact of DRECE treatment on the subsequent formation of zinc coating through hot-dip galvanizing. The primary focus was to evaluate the influence of the introduced changes on the resulting corrosion resistance, which was analyzed electrochemically using the Tafel extrapolation method. The findings can be summarized as follows:

- The DRECE method showcases its potential to enhance the hardness of the non-alloy structural steel under examination;

- It is established that applying the DRECE method does not diminish the thickness of the zinc coating of the steel specimens;
- Tafel extrapolation confirmed that the DRECE method has no effect on the corrosion resistance of the tested steel specimens;
- According to Tafel extrapolation, the impact of the DRECE method on corrosion resistance is negligible. Conversely, the hot-dip galvanizing of steel significantly enhances its corrosion resistance, which was demonstrated by a 3–4 fold reduction in the corrosion rate and a drop in the corrosion potential value;
- The influence of the electrolyte used is minimal in the case of the 0.1 M HCl and H_2SO_4 solution. However, when a 5% NaCl solution is employed, the corrosion rate experienced a significant decrease.

The DRECE method is a valuable tool that refines steel by increasing its yield strength and tensile strength in addition to its hardness while reducing its elongation. To the best of our knowledge, the coupling of the DRECE method with hot-dip galvanizing has not yet been published. This combination not only introduces the aforementioned properties but also improves the corrosion resistance of the treated steel. The potential applications of this method are vast, from the construction to the automotive industries, thereby significantly expanding its potential impact.

Author Contributions: Conceptualization, J.V.; methodology, J.V., V.N. and P.V.; validation, J.V., V.N. and P.V.; formal analysis, J.V. and V.N.; investigation, J.V., V.N. and P.V.; resources, J.V., V.N. and P.V.; data curation, J.V., V.N. and P.V.; writing—original draft preparation, J.V., V.N. and P.V.; writing—review and editing, J.V. and V.N. All authors have read and agreed to the published version of the manuscript.

Funding: This paper was created as part of the project No. CZ.02.01.01/00/22_008/0004631 Materials and technologies for sustainable development within the Jan Amos Komensky, Operational Program financed by the European Union and from the state budget of the Czech Republic and with the contribution of the student grant competition projects SP2024/062 and SP2024/025.

Institutional Review Board Statement: Not applicable.

Informed Consent Statement: Not applicable.

Data Availability Statement: The original data presented in the study are openly available in ZENODO at https://zenodo.org/records/13364426, accessed on 27 August 2024.

Conflicts of Interest: The authors declare no conflicts of interest.

References

1. Toth, L.S.; Gu, C. Ultrafine-Grain Metals by Severe Plastic Deformation. *Mater. Charact.* **2014**, *92*, 1–14. [CrossRef]
2. Valiev, R.Z.; Islamgaliev, R.K.; Alexandrov, I.V. Bulk Nanostructured Materials from Severe Plastic Deformation. *Prog. Mater. Sci.* **2000**, *45*, 103–189. [CrossRef]
3. Zrnik, J.; Dobatkin, S.V.; Mamuzič, I. Processing of Metals by Severe Plastic Deformation (SPD)—Structure and Mechanical Properties Respond. *Metalurgija* **2008**, *47*, 211–216.
4. Rusz, S.; Čízek, L.; Michenka, V.; Dutkiewicz, J.; Salajka, M.; Hilšer, O.; Tylšar, S.; Kedroň, J.; Klos, M. New Type of Device for Achievement of Grain Refinement in Metal Strip. *Arch. Mater. Sci. Eng.* **2014**, *69*, 38–44. [CrossRef]
5. Rusz, S.; Klyszewski, A.; Salajka, M.; Hilser, O.; Cizek, L.; Klos, M. Possibilities of Application Methods DRECE in Forming of Non-Ferrous Metals. *Arch. Metall. Mater.* **2015**, *60*, 3011–3016. [CrossRef]
6. Rusz, S.; Hilser, O.; Ochodek, V.; Cada, R.; Svec, J.; Szkandera, P. Influence of SPD Process on Low-Carbon Steel Mechanical Properties. *MM Sci. J.* **2019**, *2019*, 2910–2914. [CrossRef]
7. Rusz, S.; Kraus, M.; Hilšer, O.; Švec, J.; Kedroň, J.; Čížek, L.; Donič, T.; Tański, T.; Krejčí, L. Increasing the Quality of DC01 Steel by Drece Method. In Proceedings of the METAL 2017-26th International Conference on Metallurgy and Materials, Conference Proceedings, Brno, Czech Republic, 24–26 May 2017.
8. Švec, J.; Rusz, S.; Hilšer, O.; Ochodek, V.; Krejčí, L. Mechanical Properties of Steel DC01 Formed by Drece Method. *Trans. VŠB-Tech. Univ. Ostrav. Mech. Ser.* **2017**, *63*, 63–72. [CrossRef] [PubMed]
9. Jabłońska, M.B.; Kowalczyk, K.; Tkocz, M.; Bulzak, T.; Bednarczyk, I.; Rusz, S. Dual Rolls Equal Channel Extrusion as Unconventional SPD Process of the Ultralow-Carbon Steel: Finite Element Simulation, Experimental Investigations and Microstructural Analysis. *Arch. Civil. Mech. Eng.* **2021**, *21*, 1–11. [CrossRef]

10. Kowalczyk, K.; Jabłońska, M.; Rusz, S.; Bednarczyk, I. Influence of the Drece Process of Severe Plastic Deformation on the Mechanical Properties of the Ultra-Low Carbon Interstitial Free Steel. *Arch. Metall. Mater.* **2018**, *63*, 2095–2100. [CrossRef]
11. Marder, A.R. Metallurgy of Zinc-Coated Steel. *Prog. Mater. Sci.* **2000**, *45*, 191–271. [CrossRef]
12. Kania, H. Structure and Corrosion Resistance of Coatings Obtained by the Batch Double Hot Dip Method in Eutectoid ZnAl Bath with the Addition of Mg and Si. *Coatings* **2022**, *12*, 1207. [CrossRef]
13. Bhat, R.S.; Balakrishna, M.K.; Parthasarathy, P.; Hegde, A.C. Structural Properties of Zn-Fe Alloy Coatings and Their Corrosion Resistance. *Coatings* **2023**, *13*, 772. [CrossRef]
14. Kania, H.; Mendala, J.; Kozuba, J.; Saternus, M. Development of Bath Chemical Composition for Batch Hot-Dip Galvanizing—A Review. *Materials* **2020**, *13*, 4168. [CrossRef]
15. Bellini, C.; Di Cocco, V.; Iacoviello, F.; Mocanu, L.P. Impact of Copper, Tin and Titanium Addition on Bending-Induced Damage of Intermetallic Phases in Hot Dip Galvanizing. *Metals* **2022**, *12*, 2035. [CrossRef]
16. Kreislova, K.; Knotkova, D. The Results of 45 Years of Atmospheric Corrosion Study in the Czech Republic. *Materials* **2017**, *10*, 394. [CrossRef]
17. *EN ISO 1461*; Hot Dip Galvanized Coatings on Fabricated Iron and Steel Articles Specifications and Test Methods. ISO: Geneva, Switzerland, 2022.
18. *ISO 14713-2*; Zinc Coatings—Guidelines and Recommendations for the Protection against Corrosion of Iron and Steel in Structures—Part 2: Hot Dip Galvanizing. ISO: Geneva, Switzerland, 2019.
19. Sánchez, C.; Bustos, O.; Artigas, A.; Bruna, H. Silicon Effect and Microstructural Evolution of Hot Dip Galvanized Coating of Structural Steels. *Metals* **2023**, *13*, 1892. [CrossRef]
20. Vontorová, J.; Mohyla, P. Use of GDOES Method for Evaluation of the Quality and Thickness of Hot Dip Galvanised Coating. *Trans. IMF* **2018**, *96*, 313–318. [CrossRef]
21. Kuklík, V.; Kudláček, J. *Hot-Dip Galvanizing of Steel Structures*; Elsevier, Butterworth-Heinemann: Amsterdam, The Netherlands, 2016; ISBN 9780081007532.
22. Xu, W.; Wang, Y.; Yang, H. Analysis of Measurement Accuracy of Zn–Fe Alloy Coatings under Two Take-off Angles. *Vacuum* **2022**, *205*, 111456. [CrossRef]
23. Vontorová, J.; Mohyla, P.; Kreislová, K. Quality of Zinc Coating Formed on Structural Steel by Hot-Dip Galvanizing after Surface Contamination. *Coatings* **2024**, *14*, 493. [CrossRef]
24. Housecroft, C.E.; Sharpe, A.G. *Inorganic Chemistry*, 4th ed.; Pearson Education Limited: Harlow, UK, 2012; ISBN 978-0-273-74275-3.
25. Gellings, P.J.; Gierman, G.; Koster, D.; Kuit, J. Synthesis and Characterization of Homogeneous Intermetallic Fe-Zn Compounds. *Int. J. Mater. Res.* **1980**, *71*, 70–75. [CrossRef]
26. Belin, R.; Tillard, M.; Monconduit, L. Redetermination of the Iron-Zinc Phase FeZn13. *Acta Crystallogr. C* **2000**, *56*, 267–268. [CrossRef] [PubMed]
27. Okamoto, N.L.; Kashioka, D.; Inomoto, M.; Inui, H.; Takebayashi, H.; Yamaguchi, S. Compression Deformability of Γ and ζ Fe–Zn Intermetallics to Mitigate Detachment of Brittle Intermetallic Coating of Galvannealed Steels. *Scr. Mater.* **2013**, *69*, 307–310. [CrossRef]
28. Okamoto, N.L.; Inomoto, M.; Adachi, H.; Takebayashi, H.; Inui, H. Micropillar Compression Deformation of Single Crystals of the Intermetallic Compound ς-FeZn13. *Acta Mater.* **2014**, *65*, 229–239. [CrossRef]
29. Okamoto, N.L.; Tanaka, K.; Yasuhara, A.; Inui, H. Structure Refinement of the Δ1p Phase in the Fe-Zn System by Single-Crystal X-Ray Diffraction Combined with Scanning Transmission Electron Microscopy. *Acta Crystallogr. B Struct. Sci. Cryst. Eng. Mater.* **2014**, *70*, 275–282. [CrossRef]
30. Okamoto, N.L.; Yasuhara, A.; Inui, H. Order–Disorder Structure of the Δ1k Phase in the Fe–Zn System Determined by Scanning Transmission Electron Microscopy. *Acta Mater.* **2014**, *81*, 345–357. [CrossRef]
31. Hu, X.; Watanabe, T. Relationship between the Crystallographic Structure of Electrodeposited Fe-Zn Alloy Film and Its Thermal Equilibrium Diagram. *Mater. Trans.* **2001**, *42*, 1969–1976. [CrossRef]
32. Yu, J.; Liu, J.; Zhou, W.; Zhang, J.; Wu, J. Cross-Sectional TEM Observation of Iron–Zinc Intermetallic Γ and Γ1 Phases in Commercial Galvannealed IF Steel Sheets. *Mater. Des.* **2007**, *28*, 249–253. [CrossRef]
33. Hu, X.; Zhou, X. Effects of Steel Coatings Microstructure on Weldability in Resistance Spot Welding of Galvannealed Steel Sheets. *Adv. Mater. Res.* **2010**, *139–141*, 610–613. [CrossRef]
34. Weiss, Z. Emission Yields and the Standard Model in Glow Discharge Optical Emission Spectroscopy: Links to the Underlying Physics and Analytical Interpretation of the Experimental Data. *Spectrochim. Acta Part. B At. Spectrosc.* **2006**, *61*, 121–133. [CrossRef]
35. Priamushko, T.S.; Mikhaylov, A.A.; Babikhina, M.N.; Kudiiarov, V.N.; Laptev, R.S. Glow Discharge Optical Emission Spectrometer Calibration Using Hydrogenated Zr-2.5Nb Alloy Standard Samples. *Metals* **2018**, *8*, 372. [CrossRef]
36. Veverka, J.; Vilémová, M.; Chlup, Z.; Hadraba, H.; Lukáč, F.; Csáki, Š.; Matějíček, J.; Vontorová, J.; Chráska, T. Evolution of Carbon and Oxygen Concentration in Tungsten Prepared by Field Assisted Sintering and Its Effect on Ductility. *Int. J. Refract. Met. Hard Mater.* **2021**, *97*, 105499. [CrossRef]
37. Drápala, J.; Brožová, S.; Szurman, I.; Konečná, K.; Kostiuková, G.; Vontorová, J.; Jonšta, P.; Sobotková, K. Influence of Selected Rare Earth Metals on Structural Characteristics of 42CrMo4 Steel. *Metalurgija* **2016**, *55*, 757–760.
38. *ISO 643:2024*; Steels—Micrographic Determination of the Apparent Grain Size. ISO: Geneva, Switzerland, 2024.

39. *ISO 6507-1:2023*; Metallic Materials—Vickers Hardness Test, Part 1: Test Method. ISO: Geneva, Switzerland, 2023; p. 35.
40. Esmailzadeh, S.; Aliofkhazraei, M.; Sarlak, H. Interpretation of Cyclic Potentiodynamic Polarization Test Results for Study of Corrosion Behavior of Metals: A Review. *Prot. Met. Phys. Chem. Surf.* **2018**, *54.*, 976–989. [CrossRef]
41. Zhang, X.L.; Jiang, Z.H.; Yao, Z.P.; Song, Y.; Wu, Z.D. Effects of Scan Rate on the Potentiodynamic Polarization Curve Obtained to Determine the Tafel Slopes and Corrosion Current Density. *Corros. Sci.* **2009**, *51*, 581–587. [CrossRef]
42. Duarte, T.; Meyer, Y.A.; Osório, W.R. The Holes of Zn Phosphate and Hot Dip Galvanizing on Electrochemical Behaviors of Multi-Coatings on Steel Substrates. *Metals* **2022**, *12*, 863. [CrossRef]
43. Vontorová, J.; Váňová, P. Determination of Carburized Layer Thickness by GDOES Method. *AIMS Mater. Sci.* **2018**, *5*, 34–43. [CrossRef]

metals

MDPI

Article

Hot Deformation Behavior and Optimization of Processing Parameters for 4Cr16MoCu Martensitic Stainless Steel

Jiayuan Li [1,2,*], Ling Li [3], Zhongchao Wu [1,2], Tianhao Zeng [1,2] and Xiaochun Wu [1,2,*]

[1] School of Materials Science and Engineering, Shanghai University, Shanghai 200444, China
[2] State Key Laboratory of Metallurgy and Preparation of High Quality Special Steel, Shanghai University, Shanghai 200444, China
[3] School of Mechatronics Engineering, Quzhou College of Technology, Quzhou 324003, China
* Correspondence: lijiayuan_0129@163.com (J.L.); xcwu@staff.shu.edu.cn (X.W.)

Abstract: The hot deformation behavior of 4Cr16MoCu martensitic stainless steel alloyed with 1% Cu was investigated through hot compression tests at temperatures ranging from 900 to 1150 °C and strain rates of 0.001 to 1 s^{-1}. The addition of Cu is strategically employed to synergistically enhance precipitation hardening and corrosion resistance, yet its complex interplay with hot deformation mechanisms remains poorly understood, demanding systematic investigation. The results revealed a narrow forging temperature range and significant strain rate sensitivity, with deformation resistance increasing markedly at higher strain rates. An Arrhenius constitutive model incorporating a seventh-degree polynomial for strain compensation was developed to describe the flow stress dependence on deformation temperature and strain rate. The model demonstrated high accuracy, with a correlation coefficient (R^2) of 0.9917 and an average absolute relative error (AARE) of 3.8%, providing a reliable theoretical foundation for practical production applications. Furthermore, a hot processing map was constructed based on the dynamic material model (DMM), and the optimal hot working parameters were determined through microstructural analysis: an initial forging temperature of 1125 °C, a final forging temperature of 980 °C, and a strain rate of 0.1 s^{-1}. These conditions resulted in a fine and uniform grain structure, while strain rates above 0.18 s^{-1} were identified as unfavorable due to the risk of uneven deformation.

Keywords: martensitic stainless steel; hot deformation behavior; constitutive equation; processing map; microstructure

1. Introduction

4Cr16Mo martensitic stainless steel (MSS) has been extensively utilized in high-quality plastic mold steels due to its exceptional mechanical properties and corrosion resistance [1–3]. These steels typically contain alloying elements such as chromium (Cr) and molybdenum (Mo) which are integral to their performance. However, the high content of these elements, particularly in the case of 4Cr16MoCu steel, increases the deformation resistance during forging, posing significant challenges during manufacturing [4,5]. To further enhance the material's performance, the incorporation of 1wt.% Cu in this steel has been proven effective in improving its machinability and corrosion resistance. However, the introduction of copper, owing to its unique characteristics, can degrade the hot workability of the steel, necessitating the prevention of "copper embrittlement" during hot forging operations. Additionally, a certain amount of primary carbides were found in 4Cr16MoCu experimental specimens, which need to be removed through forging processes during subsequent production. Yet, there remains a dearth of research on the forging and machinability aspects of this material. Therefore, while the addition of copper to the steel enhances the application performance of the material, it also poses challenges related to its influence on the alloy's hot workability and microstructure evolution. In particular, the low impact toughness observed in large ingots of plastic mold steel represents a critical quality issue, necessitating precise control over the microstructure during free-forging to ensure final product quality [6].

One of the most critical microstructural evolution phenomena during hot deformation is discontinuous dynamic recrystallization (DRX), which is typical in high stacking-fault-energy metals such as MSS [7,8]. DRX plays a pivotal role in the microstructural evolution of alloys with low, medium, and high stacking fault energy (SFE) [9–11]. By optimizing key forging parameters, the challenges associated with coarse and uneven grain structures can be effectively addressed, thereby improving impact toughness and enhancing the overall quality of plastic mold steel blooms [12–14].

In 1969, Sellars and Jonas et al. introduced a widely recognized phenomenological model based on the Arrhenius-type equation to study material behavior during hot deformation [15,16]. This model, which establishes the relationship among stress, temperature, and strain rate, has been successfully applied to investigate the hot deformation behavior of various alloys [17,18]. The processing map (PM) technology based on the dynamic materials model (DDM) has been accepted as a powerful approach to evaluate the deformation mechanisms and optimize the process parameters of materials due to its convenience and accuracy [19,20]. It has been used in a wide range of materials including aluminum alloy, titanium, magnesium, aluminum, Ni-based alloy, as well as steels [21–23].

Despite extensive research on the hot deformation characteristics of various martensitic stainless steels [24,25], the specific hot deformation behavior of 4Cr16MoCu steel remains unexplored. To address this gap, we conducted hot compression tests on a Gleeble-3500 thermomechanical simulator under varying temperatures and strain rates. A strain-compensated constitutive equation was developed by incorporating modified flow stress values, and a hot processing map was constructed. By integrating the hot processing map with microstructural evolution analysis, the optimal hot working region was identified, providing a scientific basis for determining reasonable processing parameters in the production and manufacturing of 4Cr16MoCu MSS.

2. Experimental

Material and Experimental Procedures

The material was formulated by modifying the chemical composition of 4Cr16Mo MSS to include an additional 1 wt.% Cu, resulting in a composition of Fe-0.32C-15.94Cr-1.08Mo-0.46Si-0.60Mn-1.07Cu-0.88Ni. The incorporation of Ni, a pivotal constituent in the majority of Cu-bearing steels, is crucial due to its efficacy in mitigating hot shortness during the forging operation [26,27].

In accordance with the Gleeble hot compression specimen standards, 4Cr16MoCu steel was machined into cylindrical samples with dimensions of Φ8 mm × 12 mm using wire electrical discharge machining technology. The experiments were conducted on a Gleeble-3500 testing machine with varying parameters; specifically, the hot compression deformation temperatures were set at 900 °C, 950 °C, 1000 °C, 1050 °C, 1100 °C, and 1150 °C, and the constant strain rates were 0.001 s^{-1}, 0.01 s^{-1}, 0.1 s^{-1}, and 1 s^{-1}. The detailed process flow for the thermal compression of the specimens is illustrated in Figure 1. Initially, the specimens were heated to 1200 and homogenized for 300 s. Subsequently, they were cooled to the respective deformation temperatures, followed by a 60 s soak time prior to compression to eliminate internal temperature gradients, thereby achieving uniform temperature and microstructure distribution. Upon deforming the specimens to a strain of 0.6, water quenching was applied to rapidly cool the specimens, preserving the deformed microstructure. To obtain the microstructure of the material after hot compression, the deformed and water-quenched samples were sectioned along the axial direction. The cut surfaces were ground, polished, and etched using an aqueous solution of HCl and HNO_3. Subsequently, these etched surfaces were observed under a metallographic optical microscope.

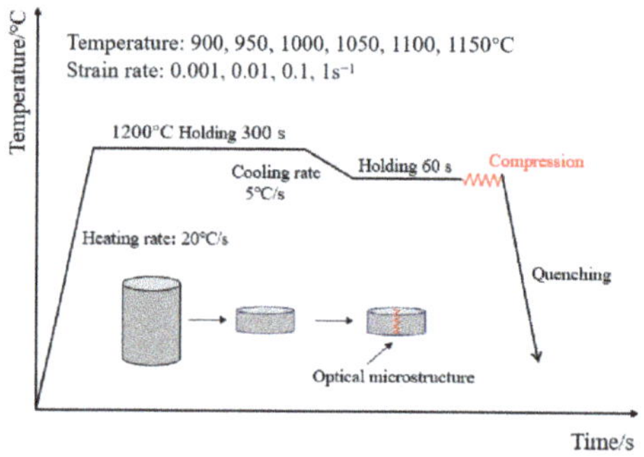

Figure 1. Schematic illustration of the hot compression test process.

3. Results and Discussion

3.1. Hot Deformation Flow Curve

The variation in flow stress during hot compression is governed by the interplay between work hardening and softening mechanisms. Figure 2 illustrates the representative true stress–strain curves of 4Cr16MoCu MSS under different deformation conditions. As shown in the figure, the flow stress increases with decreasing temperature and increasing strain rate. During the initial stage of hot compression, a substantial generation of dislocations occurs within the alloy, leading to work hardening and a rapid increase in stress with strain [28]. Subsequently, a balance is achieved between dynamic softening and work hardening, resulting in a steady-state flow stress [29]. At strain rates of 0.001 s^{-1} and 0.01 s^{-1}, the flow stress exhibits a relatively flat trend during the steady-state deformation stage. This phenomenon is attributed to the enhanced thermal activation energy at lower strain rates, which promotes recovery softening and counteracts the work hardening effect [30]. In contrast, at higher strain rates of 0.1 s^{-1} and 1 s^{-1}, the flow stress continues to increase gradually even after reaching the steady-state deformation stage. This behavior is caused by the accelerated generation of dislocations at faster compression rates, which intensifies the work hardening effect [31,32].

3.2. Constitutive Equation and Hot Deformation Activation Energy

3.2.1. Establishment of Constitutive Equation

During thermal deformation, the primary influencing factors on the true stress of metallic materials are the deformation rate and deformation temperature [33]. In this study, the Arrhenius constitutive model is employed to establish the following equation:

$$\dot{\varepsilon} = AF(\sigma)exp\left(-\frac{Q}{RT}\right) \tag{1}$$

wherein, F(σ) is a function of stress, and this expression can be further divided into three forms: the form for low stress levels ($\alpha\sigma < 0.8$), the form for high stress levels ($\alpha\sigma > 1.2$), and the form for all stress levels after the coefficients of the Arrhenius equation are modified and optimized by Sellar based on the hyperbolic sine function.

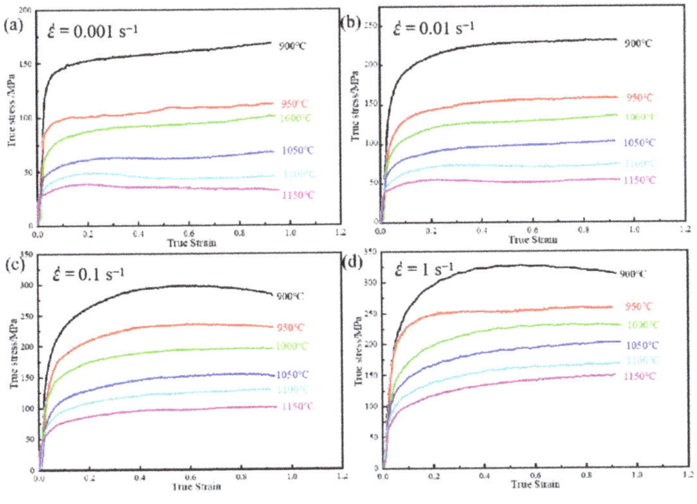

Figure 2. True stress-true strain curves of 4Cr16MoCu at different temperatures and strain rates: **(a)** $\dot{\varepsilon} = 0.001\ s^{-1}$, **(b)** $\dot{\varepsilon} = 0.01\ s^{-1}$, **(c)** $\dot{\varepsilon} = 0.1\ s^{-1}$, **(d)** $\dot{\varepsilon} = 1\ s^{-1}$.

(1) Form for low stress levels ($\alpha\sigma < 0.8$):

$$\dot{\varepsilon} = A_1\sigma^{n_1}exp\left(-\frac{Q}{RT}\right) \tag{2}$$

(2) Form for high stress levels ($\alpha\sigma > 1.2$):

$$\dot{\varepsilon} = exp(\beta\sigma)exp\left(-\frac{Q}{RT}\right) \tag{3}$$

(3) The fully stress-level form after Sellars' modification and optimization of the coefficients in the Arrhenius equation based on the hyperbolic sine function:

$$\dot{\varepsilon} = A[sinh(\alpha\sigma)]^n exp\left(-\frac{Q}{RT}\right) \tag{4}$$

In the equation, $\dot{\varepsilon}$ represents the strain rate, with units of s^{-1}; σ denotes the flow stress corresponding to a specific strain, with units of MPa; n is the stress exponent; α is the stress level parameter, obtained through the formula $\alpha = \beta/n_1$; Q represents the deformation activation energy, with units of KJ/mol; R is the ideal gas constant, typically taken as 8.314 J/(mol·K); T is the absolute temperature, with units of K; A, A_1, A_2, β, and n_1 are all material constants, which can be obtained through fitting.

Taking the characteristic point with a true strain of 0.3 as an example, the constitutive equation at this true strain value (0.3) is constructed. By taking the logarithm of both sides of Equations (2) and (3) and rearranging, we obtain

$$\ln\dot{\varepsilon} = \ln A_1 - \frac{Q}{RT} + \ln\sigma \tag{5}$$

$$\ln\dot{\varepsilon} = \ln A_2 - \frac{Q}{RT} + \beta\sigma \tag{6}$$

Substitute the strain rate $\dot{\varepsilon}$ and stress σ corresponding to the strain of 0.3 into Equations (5) and (6), and establish the relationship plots of $\ln\dot{\varepsilon} - \ln\sigma$ and $\ln\dot{\varepsilon} - \sigma$ as shown in Figures 3a and 3b, respectively, with linear fitting applied. The average values of the slopes calculated are designated as n_1 and β, yielding n_1 = 6.93758 and β = 0.055773333. Subsequently, the stress level parameter α for 4Cr16MoCu steel is obtained as 0.008039. By taking the logarithm of both sides of Equation (4) simultaneously, we derive

$$ln[sinh(\alpha\sigma)] = -\frac{1}{n}lnA + \frac{1}{n}\frac{Q}{RT} + ln\dot{\varepsilon} \quad (7)$$

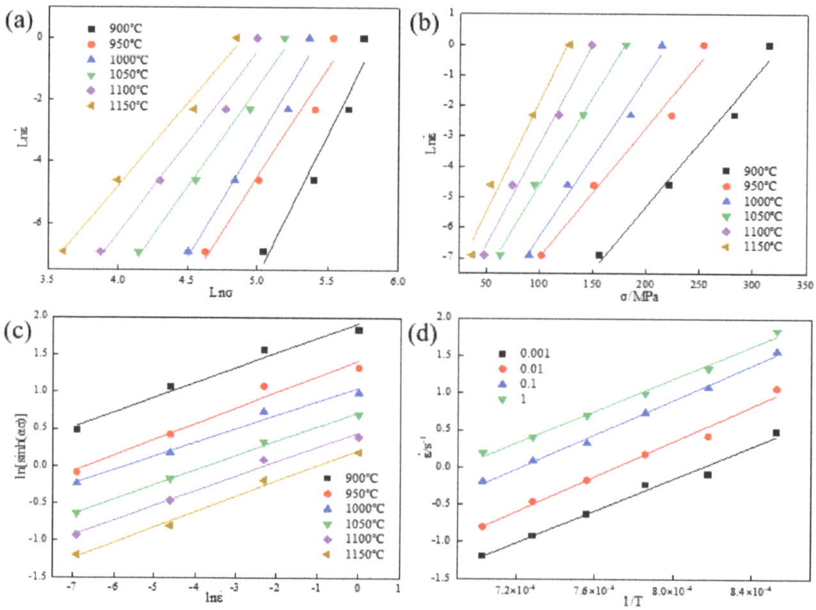

Figure 3. Linear relationships between heat distortion parameters of 4Cr16MoCu steel at a true strain of 0.3: (**a**) $\ln\dot{\varepsilon} - \ln\sigma$, (**b**) $\ln\dot{\varepsilon} - \sigma$, (**c**) $\ln[\sinh(\alpha\sigma)] - \ln\dot{\varepsilon}$, (**d**) $\ln[\sinh(\alpha\sigma)] - 1/T$.

Next, substitute the $\dot{\varepsilon}$ and σ values corresponding to the true strain of 0.3 into Equation (7). Scatter plots for $\ln[\sinh(\alpha\sigma)] - \ln\dot{\varepsilon}$ and $\ln[\sinh(\alpha\sigma)] - 1/T$ were then constructed separately and subjected to linear fitting, as illustrated in Figure 3c,d. Subsequently, the slopes of the fitted lines were calculated and averaged to obtain the stress exponent n = 5.0255 and the thermal deformation activation energy Q = 469.7088 KJ/mol for 4Cr16MoCu steel.

Zener and Hollomon introduced the deformation rate Z factor for temperature compensation [10], which quantifies the influence of deformation temperature and strain rate on the hot deformation behavior of metallic materials:

$$Z = \dot{\varepsilon}\exp\left(\frac{Q}{RT}\right) = A[\mathrm{Sinh}(\alpha\sigma)]^n \quad (8)$$

Taking the logarithm of Equation (8) yields

$$\ln[\sinh(\alpha\sigma)] = -\frac{1}{n}\ln A + \frac{1}{n}\frac{Q}{RT} + \frac{1}{n}\ln\dot{\varepsilon} \quad (9)$$

Substituting the experimental results into Equation (9), a relationship graph of $\ln Z - \ln[\sinh(\alpha\sigma)]$ is established as shown in Figure 4. Through linear fitting, the intercept lnA = 38.95096, which determines A = 8.2449 × 10^{16}. From Figure 4, the linear correlation coefficient R^2 for $\ln Z - \ln[\sinh(\alpha\sigma)]$ is 0.98844, indicating a high degree of linear fit between the two variables and thus a high level of accuracy.

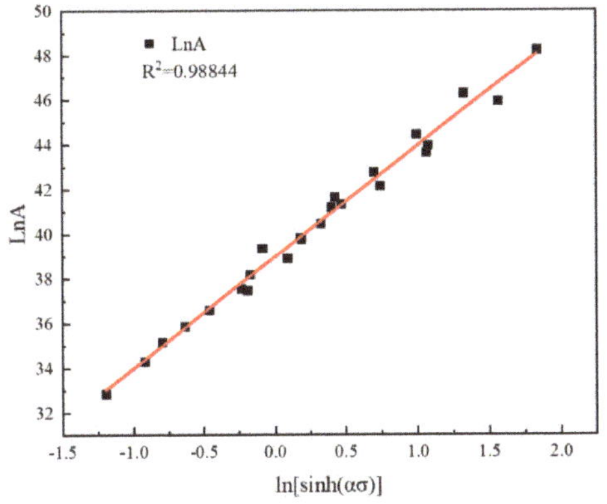

Figure 4. Relationship between peak stress and the Zener–Hollomon parameter.

In summary, substituting the calculated parameters α, n, Q, and A into Equation (4) yields the constitutive equation for 4Cr16MoCu steel at a true strain of 0.3 as follows:

$$\dot{\varepsilon} = 8.2449 \times 10^{16}[\sinh(0.0081\sigma)]^{5.0255}\exp\left(-\frac{469.7088}{RT}\right) \quad (10)$$

3.2.2. Establishment of Strain-Compensated Model

The flow stress under a single strain can be predicted through the classical Arrhenius model. However, this model neglects the influence of strain on the flow stress. Therefore, a strain-compensation modification is applied to the classical Arrhenius model to establish a relationship between strain (ε) and flow stress (σ). Based on the true stress-true strain data of 4Cr16MoCu steel under various conditions during hot compression tests, the parameter-solving process described earlier is repeated with a true strain interval of 0.05. The parameters (α, n, Q, A) of 4Cr16MoCu steel under strain conditions ranging from 0.05 to 0.9 are determined separately. The calculation results are shown in Table 1.

Table 1. Material parameters of the constitutive equations with different strain variables.

True Strain	α	n	Q (KJ/mol)	ln$A$
0.05	0.0107	6.4472	576.2463	48.6392
0.1	0.0091	5.8427	532.3549	44.8246
0.15	0.0085	5.5857	510.5521	42.8791
0.2	0.0085	5.2924	489.6513	40.6982
0.25	0.0082	5.1712	474.6575	39.3945
0.3	0.0081	5.0255	469.7088	38.9510
0.35	0.0080	4.8268	453.8576	38.5051
0.4	0.0078	4.8616	457.3048	37.9121
0.45	0.0077	4.8040	453.0186	37.5417
0.5	0.0077	4.8075	451.5751	37.4570
0.55	0.0076	4.7907	448.8429	37.2464
0.6	0.0075	4.8067	446.1386	37.0091
0.65	0.0075	4.8188	445.0589	36.9167
0.7	0.0075	4.7665	441.1404	36.5856
0.75	0.0074	4.7937	440.0870	36.5127
0.8	0.0073	4.8315	442.5761	36.1681
0.85	0.0073	4.9002	439.2116	36.5026
0.9	0.0073	4.9644	441.8771	36.7634

These four parameters all exhibit a nonlinear correlation with the strain variable ε. A seventh-degree polynomial, as shown in Equation (11), is employed to fit the relationship between strain and the model parameters. The results are presented in Figure 5, where the correlation coefficients R for each parameter against the strain are very close to 1. The specific values of these coefficients are detailed in Table 2.

$$\begin{cases} \alpha(\varepsilon) = B_0 + B_1\varepsilon + B_2\varepsilon^2 + B_3\varepsilon^3 + B_4\varepsilon^4 + B_5\varepsilon^5 + B_6\varepsilon^6 + B_7\varepsilon^7 \\ n(\varepsilon) = C_0 + C_1\varepsilon + C_2\varepsilon^2 + C_3\varepsilon^3 + C_4\varepsilon^4 + C_5\varepsilon^5 + C_6\varepsilon^6 + C_7\varepsilon^7 \\ Q(\varepsilon) = D_0 + D_1\varepsilon + D_2\varepsilon^2 + D_3\varepsilon^3 + D_4\varepsilon^4 + D_5\varepsilon^5 + D_6\varepsilon^6 + D_7\varepsilon^7 \\ \ln A(\varepsilon) = E_0 + E_1\varepsilon + E_2\varepsilon^2 + E_3\varepsilon^3 + E_4\varepsilon^4 + E_5\varepsilon^5 + E_6\varepsilon^6 + E_7\varepsilon^7 \end{cases} \quad (11)$$

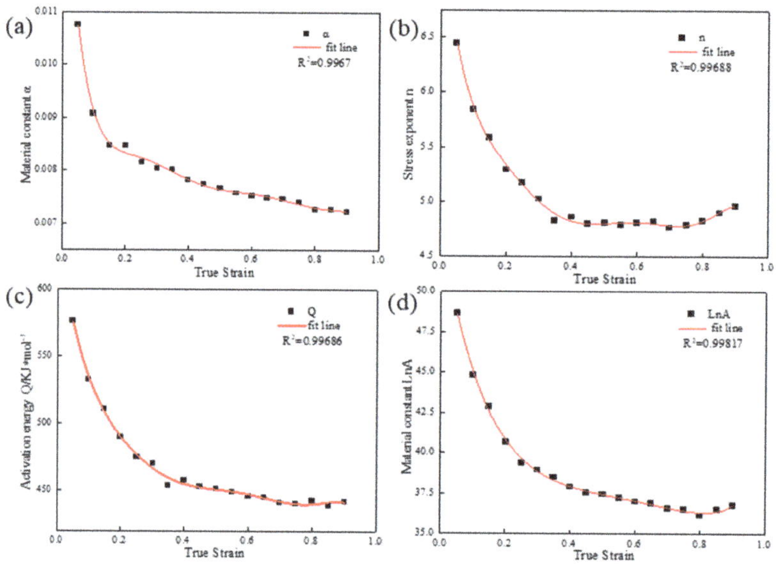

Figure 5. The relationship between equation parameters and true strain: (**a**) α, (**b**) n, (**c**) Q, (**d**) lnA.

Table 2. Seventh-degree polynomial coefficients for *α*, *n*, *Q*, ln*A* parameter.

Stress Level Parameter (*α*)	Stress Exponent (*n*)	Strain Activation Energy (*Q*)	ln*A*
$B_0 = 0.0146$	$C_0 = 7.6244$	$D_0 = 644.7489$	$E_0 = 53.6681$
$B_1 = -0.1091$	$C_1 = -33.1578$	$D_1 = -1790.6077$	$E_1 = -119.4691$
$B_2 = 0.7925$	$C_2 = 235.8938$	$D_2 = 9861.2017$	$E_2 = 399.5431$
$B_3 = -3.0455$	$C_3 = -1028.2501$	$D_3 = -37{,}700$	$E_3 = -779.8986$
$B_4 = 6.5608$	$C_4 = 2576.0709$	$D_4 = 91{,}000$	$E_4 = 933.1345$
$B_5 = -7.9708$	$C_5 = -3594.5561$	$D_5 = -127000$	$E_5 = -648.6062$
$B_6 = 5.0996$	$C_6 = 2593.1982$	$D_6 = 93{,}344.5432$	$E_6 = 190.1419$
$B_7 = -1.3358$	$C_7 = -752.7588$	$D_7 = -27{,}634.6160$	$E_7 = 11.0290$

Deriving the stress σ from Equation (4) and rearranging, the flow stress of the material can be expressed by Equation (12):

$$\sigma = \frac{1}{\alpha}\sinh^{-1}\left[\left(\frac{Z}{A}\right)^{\frac{1}{n}}\right] \quad (12)$$

Furthermore, based on the definition of the hyperbolic sine function, Equation (12) can also be represented as

$$\sigma = \frac{1}{\alpha}\ln\left\{\left(\frac{Z}{A}\right)^{\frac{1}{n}} + \left[\left(\frac{Z}{A}\right)^{\frac{2}{n}} + 1\right]^{\frac{1}{2}}\right\} \tag{13}$$

By substituting the strain-dependent flow stress parameter relationship (Equation (11)) obtained from the above calculations into Equation (13), the high-temperature flow stress model for 4Cr16MoCu steel under different deformation temperatures and train rates is obtained, as shown in Equation (14):

$$\begin{cases} \sigma = \frac{1}{\alpha(\varepsilon)}\ln\left\{\left(\frac{Z(\varepsilon)}{A(\varepsilon)}\right)^{\frac{1}{n(\varepsilon)}} + \left[\left(\frac{Z(\varepsilon)}{A(\varepsilon)}\right)^{\frac{2}{n(\varepsilon)}} + 1\right]^{\frac{1}{2}}\right\} \\ Z(\varepsilon) = \dot{\varepsilon}\exp\left(\frac{Q(\varepsilon)}{RT}\right) \end{cases} \tag{14}$$

3.2.3. Verification of the Developed Constitutive Model

Using Equation (14), the stress values corresponding to different true strains, deformation temperatures, and strain rates for 4Cr16MoCu steel were calculated. The actual experimental values were compared with the model-calculated values to validate the model; the results are shown in Figure 6, in the figure, the solid line corresponds to the experimental results, whereas the colored dots represent the values predicted by the model. From Figure 6b, it can be observed that there is a slight discrepancy between the flow stress data and the actual values for the condition of 900 °C + 0.01 s^{-1}. However, the errors between the calculated values and the experimental values under the remaining deformation conditions are relatively small. This indicates that the predicted values calculated by the equation are generally in good agreement with the actual stress–strain curves.

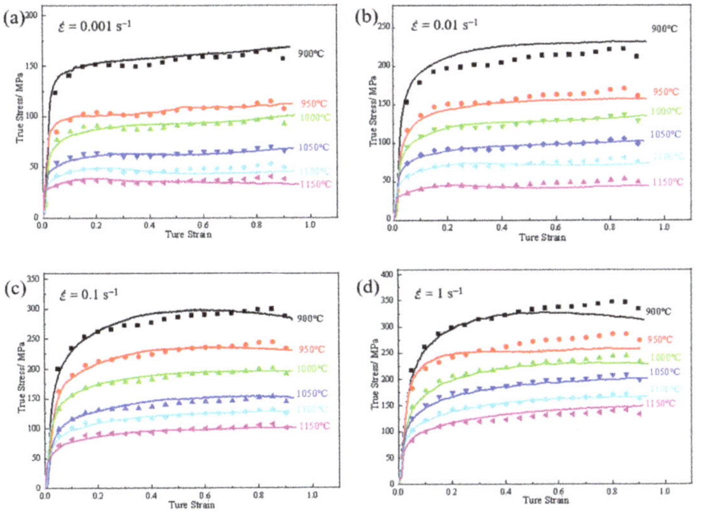

Figure 6. Comparison of experimental and calculated values of rheological stress in 4Cr16MoCu steel at different strain rates: **(a)** $\dot{\varepsilon}$ = 0.001 s^{-1}, **(b)** $\dot{\varepsilon}$ = 0.01 s^{-1}, **(c)** $\dot{\varepsilon}$ = 0.1 s^{-1}, **(d)** $\dot{\varepsilon}$ = 1 s^{-1}.

To quantitatively assess the accuracy of the present equations, as illustrated in Equations (15) and (16), the correlation coefficient (R) and the Average Relative Error (AARE) are introduced for quantitative analysis of the model [34,35].

$$R = \frac{\sum_{i=1}^{N}(E_i - \overline{E})(P_i - \overline{P})}{\sqrt{\sum_{i=1}^{N}(E_i - \overline{E})^2 - \sum_{i=1}^{N}(P_i - \overline{P})^2}} \tag{15}$$

$$AARE = \frac{1}{N}\sum_{i=1}^{N}\left|\frac{E_i - P_i}{E_i}\right| \times 100\% \tag{16}$$

In these equations, E_i, $\overline{E}$ represent the experimental stress values and their average, respectively; P_i, $\overline{P}$ denote the stress values predicted by the model and their average, respectively; N stands for the number of experimental data points; and i indexes the experimental sequences.

As shown in Figure 7, the solid line represents the best-fit curve between the experimental values and the model-predicted values. The results indicate that the correlation coefficient and the average relative error are 0.9917 and 3.8%, indicating a relatively small overall error. Therefore, this strain-compensated constitutive equation can effectively predict the flow stress during the hot deformation process and provide a theoretical basis for formulating a reasonable forming process for large-scale forgings.

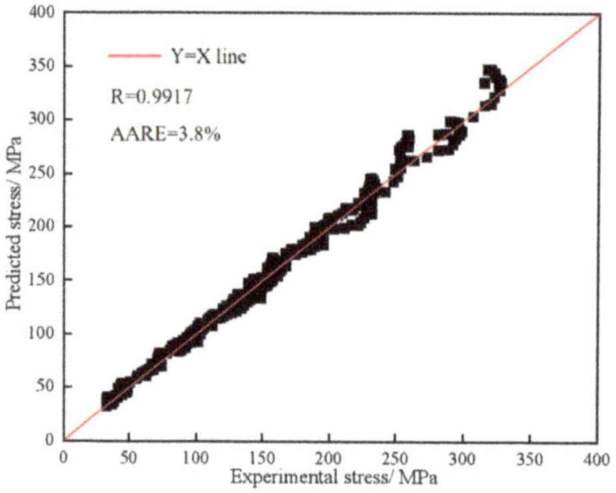

Figure 7. Correlation between the predicted value of the model and the experimental value.

3.3. *Effects of Strain on Processing Maps and Microstructural Evolution*

The thermomechanical processing map is established based on the dynamic materials model (DMM) proposed by Prasad et al. [36], which allows for the determination of safe and unstable regions during the hot working of materials, ultimately to determine the optimum hot deformation condition for materials [37,38].

3.3.1. Establishment of Power Dissipation Maps

This model posits that the total energy P absorbed by a hot-worked workpiece instantaneously during deformation consists of a dissipative component G and a co-dissipative component J according to Equation (17).

$$P = \sigma\dot{\varepsilon} = G + J = \int_0^{\dot{\varepsilon}} \sigma d\dot{\varepsilon} + \int_0^{\sigma} \dot{\varepsilon} d\sigma \tag{17}$$

Once the temperature and strain during the hot working of the workpiece are determined, the relationship between the stress σ experienced by the workpiece and the strain rate can be expressed by Equation (18):

$$\sigma = K\dot{\varepsilon}^{m} \tag{18}$$

In this equation, K represents the flow stress at a strain rate of 1, m is the strain rate sensitivity exponent, which governs the power dissipation G and the dissipation covariant J during the thermal deformation process of the material. The value of m can be obtained through curve fitting of the lnσ-lnε relationship, with its specific definition given by

$$m = \left[\frac{\partial J}{\partial G}\right]_{\varepsilon,T} = \left.\frac{\partial \ln\sigma}{\partial \dot{\varepsilon}}\right|_{\varepsilon,T} \tag{19}$$

H. Ziegler [39] pointed out that for the steady-state rheological behavior of viscoplastic solids, the value of m ranges between 0 and 1, and a higher value of m indicates a greater energy requirement for microstructural evolution. When m is constant, the dissipation covariant J can be simplified from Equations (3)–(17) as follows:

$$J = \int_0^{\sigma} \dot{\varepsilon}\, d\sigma = \int_0^{\sigma} \left(\frac{\dot{\sigma}}{K}\right)^{\frac{1}{m}} d\sigma = \frac{m\sigma\dot{\varepsilon}}{1+m} \tag{20}$$

For nonlinear dissipative systems, due to the continuous nonlinear variation of J, for the sake of computational convenience, it is compared with the ideal linear dissipative factor J_{max} to generate a dimensionless parameter, namely η, which represents the energy dissipation efficiency arising from microstructural evolution:

$$\eta = \frac{J}{J_{max}} = \frac{2m}{m+1} \tag{21}$$

In the equation, η is termed the energy dissipation efficiency factor; J_{max} represents the maximum dissipation covariant when m = 1, at which point the material is in an ideal linear dissipative state. The distribution plot drawn by the η values obtained from different temperatures T and strain rates $\dot{\varepsilon}$ forms the power dissipation map. The microstructure evolution patterns of the workpiece during hot working deformation can be inferred from the distribution of η values in the power dissipation map.

Next, taking the strain of 0.2 as an example, the power dissipation map and instability map for 4Cr16MoCu are plotted. Initially, based on the stress–strain curves depicted in Figure 2, the stress value σ corresponding to the strain of 0.2 is extracted. A $\sigma - \ln\dot{\varepsilon}$ scatter plot is constructed and subjected to linear fitting, with the results presented in Figure 8a. It is evident from the figure that there is a good linear correlation between the two variables. Subsequently, a $\ln\sigma - \ln\dot{\varepsilon}$ curve is plotted and fitted using a cubic polynomial, as shown in Figure 8b. The cubic spline function, as expressed in Equations (3)–(22), is then solved:

$$\ln\sigma = k + b_1\ln\dot{\varepsilon} + b_2[\ln\dot{\varepsilon}]^2 + b_3[\ln\dot{\varepsilon}]^3 \tag{22}$$

The coefficients k, b_1, b_2, b_3 of the cubic function in the figure are used to calculate the values under various conditions. A $(\ln\sigma)\prime - \ln\dot{\varepsilon}$ scatter plot is constructed and fitted, with the results displayed in Figure 8c. Combining Equation (19), by taking the partial derivative of Equation (22) again, the value of m can be obtained. Further, η can be derived using Equation (21).

$$m = \frac{\partial \ln\sigma}{\partial \ln\dot{\varepsilon}} = b_1 + 2b_2\ln\dot{\varepsilon} + 3b_3(\ln\dot{\varepsilon})^2 \tag{23}$$

For subsequent comparison of hot working maps under different true strain values, similarly, the η values at strains of 0.4, 0.6, and 0.8 are calculated using the aforementioned method. The specific values are presented in Table 3.

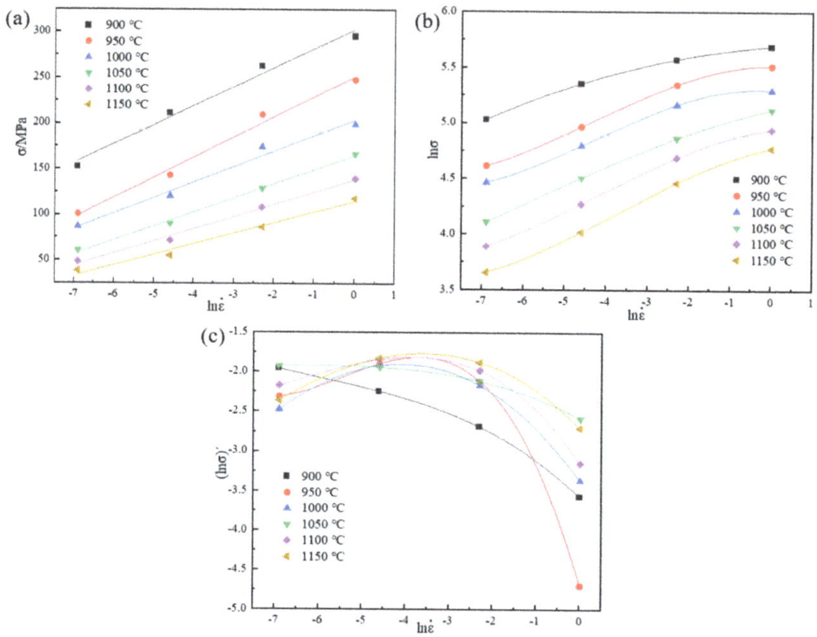

Figure 8. Fitted plots for parameters at strain 0.2: (**a**) $\sigma - \ln\dot{\varepsilon}$, (**b**) $\ln\sigma - \ln\dot{\varepsilon}$, (**c**) $(\ln\sigma)\prime - \ln\dot{\varepsilon}$.

Table 3. Energy dissipation efficiency factor values (η) under different strain conditions.

Strain	Strain Rate (s^{-1})	Energy Dissipation Efficiency Factor Values (η)					
		900 °C	950 °C	1000 °C	1050 °C	1100 °C	1150 °C
0.2	0.001	0.2827	0.1981	0.1692	0.2929	0.2283	0.1888
	0.01	0.2112	0.3005	0.2933	0.2867	0.3165	0.3221
	0.1	0.1359	0.2411	0.2287	0.2415	0.2759	0.3066
	1	0.0565	0.0183	0.0691	0.1508	0.0858	0.1335
0.4	0.001	0.2763	0.1837	0.1196	0.2931	0.2433	0.2080
	0.01	0.2242	0.3114	0.2955	0.3291	0.4097	0.4263
	0.1	0.1364	0.2299	0.2505	0.2723	0.3467	0.3982
	1	0.0022	0.1174	0.0526	0.1019	0.0104	0.0936
0.6	0.001	0.2698	0.1642	0.1234	0.3127	0.2614	0.2341
	0.01	0.2448	0.3263	0.3080	0.3372	0.4043	0.4009
	0.1	0.1543	0.2369	0.2609	0.2695	0.3398	0.4156
	1	0.0233	0.1689	0.0586	0.0844	0.0085	0.0269
0.8	0.001	0.2922	0.2022	0.1478	0.2979	0.2916	0.1014
	0.01	0.2415	0.3288	0.3023	0.3240	0.3736	0.3955
	0.1	0.1497	0.2355	0.2504	0.2635	0.3073	0.3815
	1	0.0039	0.1438	0.0491	0.0960	0.0509	0.0412

At a given strain, plotting the values of η at various temperatures and strain rates onto a contour map constitutes the power dissipation map. The power dissipation efficiency map on the $T - \lg(\dot{\varepsilon})$ plane is plotted using Origin software (OriginPro 2018C), with the results shown in Figure 9. Combining the results from Table 3, it can be observed that the variation trend of η values for 4Cr16MoCu steel under different strains tends to be consistent: the maximum η values at various strains are concentrated in the low strain rate region near 0.01~0.1 s^{-1}, with temperature ranges mainly focused on approximately 950~1000 °C and 1100~1150 °C. The combination of these two temperature ranges can be preliminarily considered as the optimal hot working region for 4Cr16MoCu steel. Comparing the η values under the same parameters among the four strains, the results indicate that the η value of 4Cr16MoCu steel also increases as the strain increases. This is due to the increasing number of dislocations within the steel as strain progresses, which subsequently increases the driving force required for dynamic recrystallization to occur within the microstructure. According to the dynamic materials model theory, an increase in the η value also represents an increase in the energy utilized for microstructural evolution.

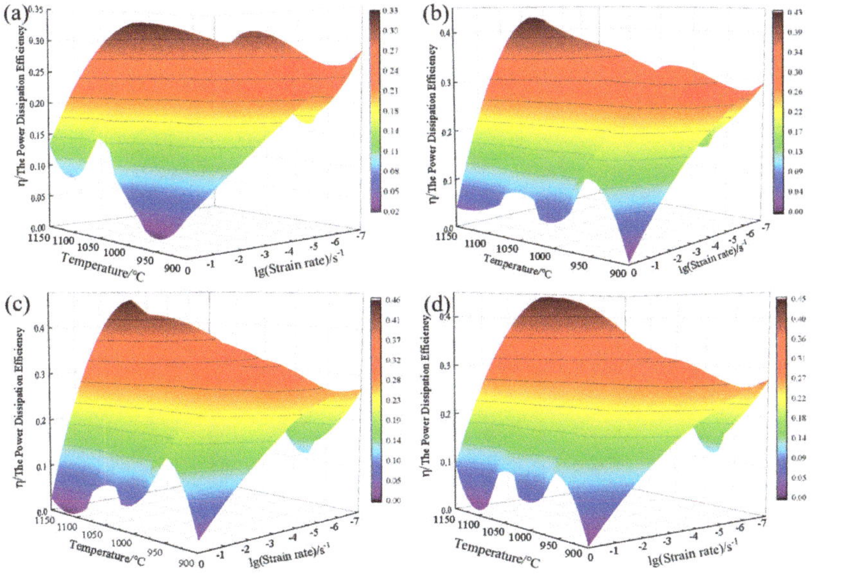

Figure 9. Power dissipation map at each strain: **(a)** ε = 0.2, **(b)** ε = 0.4, **(c)** ε = 0.6, **(d)** ε = 0.8.

3.3.2. Construction of Instability Maps

Generally, a higher power dissipation efficiency factor (η) indicates that the material consumes more energy during microstructural evolution, making the material more prone to recrystallization and suggesting that the material will exhibit better processability [40]. However, high η values can also be observed in the region of processing instability. Therefore, based on the principle of extreme values in irreversible thermodynamics of large plastic deformation, Prasad and Murthy proposed a processing instability criterion:

$$\xi(\dot{\varepsilon}) = \frac{\partial \ln\left(\frac{m}{m+1}\right)}{\partial \ln \dot{\varepsilon}} + m < 0 \tag{24}$$

The physical significance of Equation (24) is that when the entropy production rate of a system is less than the strain rate applied to the system, i.e., $\xi < 0$, plastic deformation will localize, leading to flow instability.

Similarly, by calculating the ξ values and constructing contour plots of ξ as a function of temperature and strain rate, the instability parameter region for 4Cr16MoCu steel can be determined. For a strain of 0.2, the value of $\ln(m/(m+1))$ is first calculated using the m values obtained in Equation (23) Then, a cubic spline function is used to fit the functional relationship between $\ln(m/(m+1))$ and $\ln\dot{\varepsilon}$, and the coefficients e, f, g, h are obtained through regression analysis.

$$\ln\left(\frac{m}{m+1}\right) = e + f\ln\dot{\varepsilon} + g(\ln\dot{\varepsilon})^2 + h(\ln\dot{\varepsilon})^3 \tag{25}$$

$$\xi(\dot{\varepsilon}) = \frac{\partial\ln\left(\frac{m}{m+1}\right)}{\partial\ln\dot{\varepsilon}} + m = \left(1 + 2m(\ln\dot{\varepsilon}) + 3n(\ln\dot{\varepsilon})^2\right) + m < 0 \tag{26}$$

By substituting the coefficients e, f, g, h and their corresponding m values into Equation (26), the ξ values can be obtained, as shown in Table 4.

Table 4. The corresponding ξ value under different strain conditions.

Strain	Strain Rate (s^{-1})	ξ Value					
		900 °C	950 °C	1000 °C	1050 °C	1100 °C	1150 °C
0.2	0.001	0.0291	0.1804	0.4833	0.1730	0.2901	0.4347
	0.01	−0.0202	0.3442	0.2482	0.1361	0.2703	0.3118
	0.1	−0.1928	−0.3459	−0.1740	0.0086	−0.0823	0.0042
	1	−0.4888	−1.8713	−0.7117	−0.2097	−0.7677	−0.4883
0.4	0.001	−0.2678	0.5979	0.5095	0.2009	0.2071	0.8331
	0.01	0.2020	0.1821	0.3315	0.2041	0.3250	0.5915
	0.1	−0.6080	−0.0939	−0.2128	−0.0752	−0.1673	−0.1980
	1	−2.6978	−0.0754	−1.0734	−0.6370	−1.2697	−1.5352
0.6	0.001	0.0389	0.7488	0.6622	0.1904	0.0514	0.1461
	0.01	0.0951	0.2028	0.3627	0.2163	0.5210	0.5398
	0.1	−0.3503	−0.0802	−0.1923	−0.0990	−0.4243	−0.1603
	1	−1.2738	0.0845	−0.9424	−0.7557	−2.7846	−2.1076
0.8	0.001	−0.3490	0.5778	0.6415	0.2171	0.1301	0.5128
	0.01	0.2139	0.1997	0.3574	0.2161	0.5265	0.4551
	0.1	−0.6865	−0.1153	−0.2081	−0.0617	−0.3963	−0.0376
	1	−3.0502	−0.2427	−1.0012	−0.6163	−2.6278	−0.9651

Similarly, using Origin software, different ξ values are substituted onto the $T - \lg(\dot{\varepsilon})$ plane, and contour maps are plotted to represent the instability maps of the steel under various strains, as shown in Figure 10. The negative ξ value regions in Figure 10 indicate the hot forging parameters where instability occurs during the hot working of 4Cr16MoCu steel. By comparing the instability maps at different strains, it can be seen that the instability phenomenon in 4Cr16MoCu steel is mainly controlled by the strain rate. When the strain rate exceeds 0.1 s^{-1}, the material is highly susceptible to instability. For steel grades with poor plasticity, such as this one, forging at a high deformation rate during hot forging can easily lead to instability phenomena, such as cracking.

Subsequently, contour plots are formulated utilizing the ξ values to delineate the zones within the workpiece where deformation instability manifests, termed the instability map. Through the superposition of the power dissipation map onto the instability map, the hot working map pertinent to the specified deformation conditions is derived. Figure 11 illustrates the establishment of hot working maps corresponding to distinct strain conditions, where gray-shaded regions denote instability zones, whereas the unshaded areas signify the suitable processing domains.

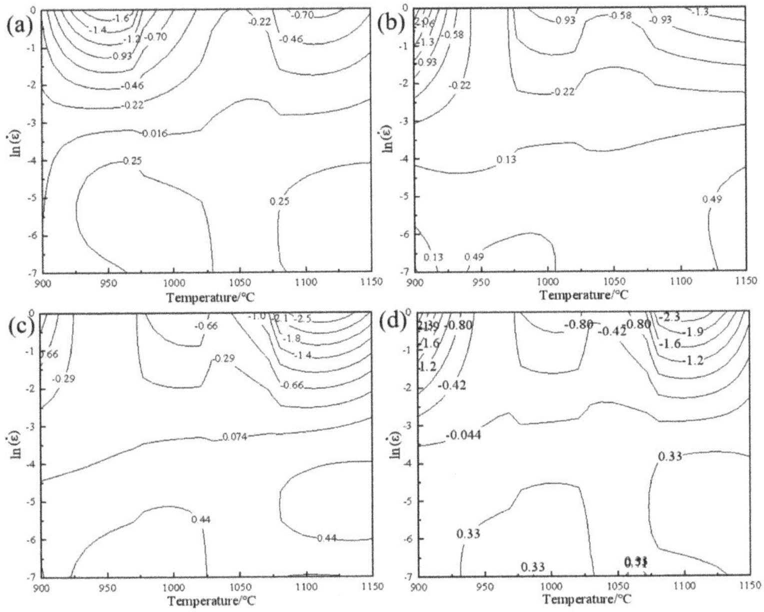

Figure 10. Instability map at each strain: (**a**) ε = 0.2, (**b**) ε = 0.4, (**c**) ε = 0.6, (**d**) ε = 0.8.

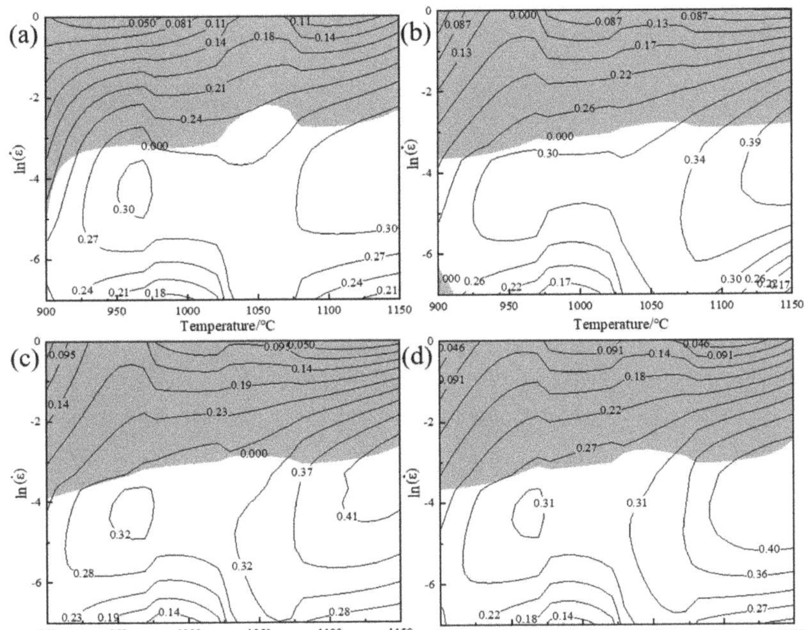

Figure 11. Hot working map with strain of (**a**) ε = 0.2, (**b**) ε = 0.4, (**c**) ε = 0.6, (**d**) ε = 0.8.

By analyzing the hot working maps across four strain levels, it is evident that for 4Cr16MoCu steel, the flow instability zone predominantly concentrates in the high strain rate region of approximately 0.01 s^{-1} and higher. The maximum permissible thermal deformation rate of the material, as deduced from Figure 11a, is approximately 0.018 s^{-1}.

Throughout the entire hot forging process, minimal variations are observed in the instability zone, suggesting inadequate plastic deformation characteristics for this steel grade. Considering the inconsistency in strain across various parts of the steel component during hot forging, along with the η values under different strain parameters depicted in the figure, the peak η value occurs in two regions: one within the range of 950–1000 °C at a strain rate of 0.01 s^{-1} with an η value of approximately 0.32, and another within the range of 1100–1150 °C at a strain rate of 0.01 s^{-1} with an η value reaching approximately 0.4. Given the practical requirements of hot forging to determine the initial and final forging temperatures, these two sets of parameters can be adopted as the suitable hot forging parameters for 4Cr16MoCu steel, namely an initial forging temperature of 1125 °C, a final forging temperature of 980 °C, and a strain rate set at 0.01 s^{-1}.

Figure 12 illustrates the microstructures of 4Cr16MoCu specimens subsequent to selecting representative hot compression parameters. Specifically, Figure 12a and b depict the microstructures of the instability zones at 900 °C + 1 s^{-1} and 1000 °C + 1 s^{-1}, respectively. Both specimens exhibit localized deformation bands internally, with grains predominantly undergoing deformation and elongation along the direction of hot compression. As temperature increases, grain size augments. The microstructure exhibits a mixed grain phenomenon attributed to incomplete recrystallization, which degrades the workpiece properties and predisposes it to instability [41]. Figure 12c,d display the microstructures of the regions at 1100 °C + 0.001 s^{-1} and 1100 °C + 0.01 s^{-1}, respectively. The results indicate the occurrence of dynamic recrystallization in both samples. However, a lower deformation rate leads to the formation of larger grains, adversely affecting the material properties. Consequently, maintaining a strain rate of 0.01 s^{-1} results in a more uniform microstructure of the workpiece.

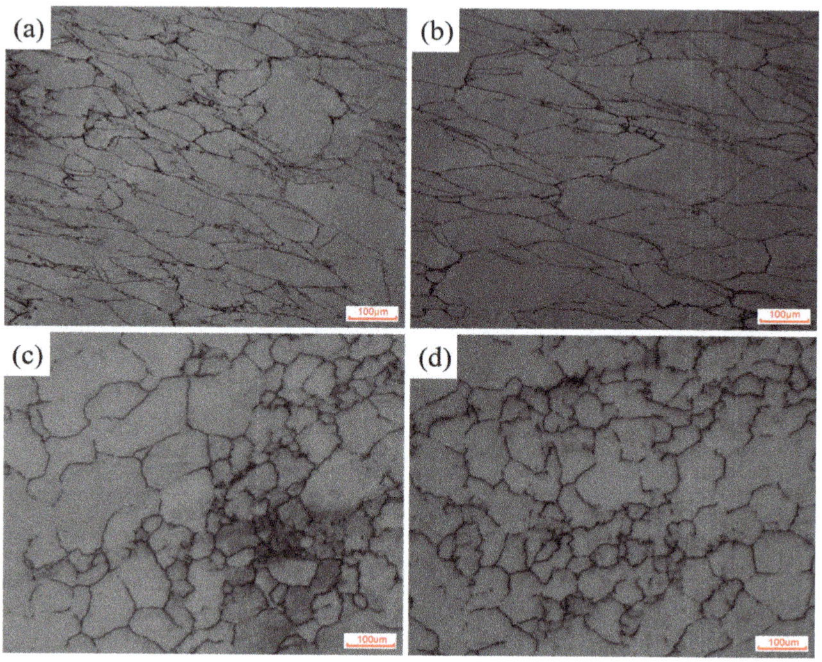

Figure 12. Microstructure of 4Cr16MoCu under different deformation conditions: (**a**) 900 °C, $\dot{\varepsilon}$ = 1 s^{-1}, (**b**) 1000 °C, $\dot{\varepsilon}$ = 1 s^{-1}, (**c**) 1100 °C, $\dot{\varepsilon}$ = 0.001 s^{-1}, (**d**) 1100 °C, $\dot{\varepsilon}$ = 0.01 s^{-1}.

4. Conclusions

1. The forging temperature range of 4Cr16MoCu stainless steel is relatively narrow. During the forging process, its deformation resistance increases significantly with the rise in deformation rate, exhibiting a pronounced sensitivity to strain rate.
2. The Arrhenius model constitutive equation was established using corrected flow stress. The dependence of flow stress on strain rate and temperature during hot difformation can be expressed as

$$\dot{\varepsilon} = 8.2449 \times 10^{16}[\sinh(0.0081\sigma)]^{5.0255} \exp\left(-\frac{469.7088}{RT}\right) \quad (27)$$

This demonstrates the reliability of the model for practical applications. The findings provide valuable insights into the hot deformation behavior of 4Cr16MoCu steel, and the established equation can serve as a theoretical foundation for optimizing processing parameters in industrial applications.

3. Based on the dynamic materials model, a hot processing map for 4Cr16MoCu steel was established, and the optimal hot working process window for 4Cr16MoCu steel was determined in conjunction with microstructural evolution. The optimal conditions include an initial forging temperature of 1125 °C, a final forging temperature of 980 °C, and a strain rate of 0.1 s^{-1}, which result in a fine and uniform grain structure. To avoid uneven deformation, high strain rate areas of 0.18 s^{-1} and above should be avoided during hot processing.

Author Contributions: J.L.: conceptualization, validation, investigation, data curation, writing—original draft preparation. L.L.: conceptualization, validation, investigation, data curation, writing—review and editing. Z.W.: software, data curation, visualization. T.Z.: methodology, formal analysis, investigation X.W.: resources, review and editing, visualization, project administration, funding acquisition. All authors have read and agreed to the published version of the manuscript.

Funding: The work is supported by the National Key Research and Development Program of Guangdong Province (Grant No. 2020B010184002).

Data Availability Statement: The original contributions presented in the study are included in the article, further inquiries can be directed to the corresponding author.

Conflicts of Interest: The authors declare no competing interests.

References

1. Bonagani, S.K.; Bathula, V.; Kain, V. Influence of tempering treatment on microstructure and pitting corrosion of 13 wt.% Cr martensitic stainless steel. *Corros. Sci.* **2018**, *131*, 340–354.
2. Anjos, A.D.; Scheuer, C.J.; Brunatto, S.F.; Cardoso, R.P. Low-temperature plasma nitrocarburizing of the AISI 420 martensitic stainless steel: Microstructure and process kinetics. *Surf. Coat. Technol.* **2015**, *275*, 51–57.
3. Zhao, Y.; Liu, W.; Fan, Y.; Zhang, T.; Dong, B.; Chen, L.; Wang, Y. Influence of Microstructure on the Corrosion Behavior of Super 13Cr Martensitic Stainless Steel under Heat Treatment. *Mater. Charact.* **2021**, *175*, 111066.
4. Islam, R.; Islam, J.; Hasan, R.; Hasan, M. Effect of Alloying Element Content, Temperature, and Strain Rate on the Mechanical Behavior of NbTiZrMoV High Entropy Alloy: A Molecular Dynamics Study. *Mater. Today Commun.* **2024**, *40*, 110071.
5. Liu, W.; Liu, W.M.; Ji, H.; Tang, X.; Wang, M.; Song, C.; Yang, X. Hot Deformation Behavior and Hot Working Map of Mn–Cr–Ni–Co Steel for Ball Mill Liner Forging Process. *J. Mater. Res. Technol.* **2024**, *30*, 5685–5700.
6. Li, X.; Duan, L.; Li, J.; Wu, X. Experimental Study and Numerical Simulation of Dynamic Recrystallization Behavior of a Micro-Alloyed Plastic Mold Steel. *Mater. Des.* **2015**, *66*, 309–320.
7. Yang, Y.; Yan, Q.Z.; Ge, C.C. Hot Deformation Behavior of Modified CNS-II F/M Steel. *J. Iron Steel Res. Int.* **2012**, *19*, 60–65.
8. Aghajani Derazkola, H.; Garcia, E.; Murillo-Marrodán, A.; Fernandez, A.C. Review on Modeling and Simulation of Dynamic Recrystallization of Martensitic Stainless Steels during Bulk Hot Deformation. *J. Mater. Res. Technol.* **2022**, *18*, 2993–3025.

9. Momeni, A.; Dehghani, K.; Heidari, M. Modeling the Flow Curve of AISI 410 Martensitic Stainless Steel. *J. Mater. Eng. Perform.* **2012**, *21*, 2238–2243.
10. Dehghan, H.; Abbasi, S.M.; Momeni, A.; Karimi Taheri, A. On the Constitutive Modeling and Microstructural Evolution of Hot Compressed A286 Iron-Base Superalloy. *J. Alloys Compd.* **2013**, *564*, 13–19.
11. Xu, Y.; Liu, J.S.; Jiao, Y.X. Hot Deformation Behavior and Dynamic Recrystallization Characteristics of 12Cr Ultra-Super-Critical Rotor Steel. *Metals Mater. Int.* **2019**, *25*, 823–837. [CrossRef]
12. Jantepa, N.; Siripath, N.; Suranuntchai, S. Predictive Modeling and Optimization of Hot Forging Parameters for AISI 1045 Ball Joints Using Taguchi Methodology and Finite Element Analysis. *Metals* **2024**, *14*, 1198. [CrossRef]
13. Fang, X.R.; Liu, L.; Lu, J.; Gao, Y. Optimization of Forging Process Parameters and Prediction Model of Residual Stress of Ti-6Al-4V Alloy. *Adv. Mater. Sci. Eng.* **2021**, *1*, 3105470. [CrossRef]
14. Kitayama, S. Technical Review on Design Optimization in Forging. *Int. J. Adv. Manuf. Technol.* **2024**, *132*, 4161–4189. [CrossRef]
15. Sellars, C.M.; McTegart, W.J. On the Mechanism of Hot Deformation. *Acta Metall.* **1966**, *14*, 1136–1143. [CrossRef]
16. Jonas, J.J.; Sellars, C.M.; Tegart, W.J.M. Strength and Structure under Hot-Working Conditions. *Metall. Rev.* **1969**, *14*, 1–24. [CrossRef]
17. Wang, Y.; Peng, J.; Zhong, L.; Pan, F. Modeling and Application of Constitutive Model Considering the Compensation of Strain During Hot Deformation. *J. Alloys Compd.* **2016**, *681*, 455–470. [CrossRef]
18. Wu, H.; Wen, S.P.; Huang, H.; Wu, X.L.; Gao, K.Y.; Wang, W.; Nie, Z.R. Hot Deformation Behavior and Constitutive Equation of a New Type Al–Zn–Mg–Er–Zr Alloy during Isothermal Compression. *Mater. Sci. Eng. A* **2016**, *651*, 415–424. [CrossRef]
19. Babu, K.A.; Mandal, S.; Athreya, C.N.; Shakthipriya, B.; Sarma, V.S. Hot Deformation Characteristics and Processing Map of a Phosphorous Modified Super Austenitic Stainless Steel. *Mater. Des.* **2017**, *115*, 262–275. [CrossRef]
20. Park, S.Y.; Kim, W.J. Difference in the Hot Compressive Behavior and Processing Maps between the As-Cast and Homogenized Al-Zn-Mg-Cu (7075) Alloys. *J. Mater. Sci. Technol.* **2016**, *32*, 660–670. [CrossRef]
21. Mukherjee, T.; Gao, M.; Palmer, T.A.; DebRoy, T. Keyhole Mode Wobble Laser Welding of a Nickel Base Superalloy—Modeling, Experiments, and Process Maps. *J. Manuf. Process.* **2023**, *106*, 465–479.
22. Fuh, Y.-K.; Shih, J.-K.; Saputro, I.E.; Chen, C.-P.; Huang, C.-F.; Ku, H.-Y.; Chan, C.-W. Preform Design with Increased Materials Utilization and Processing Map Analysis for Aluminum Hot Forging Process. *J. Manuf. Process.* **2023**, *90*, 14–27.
23. Younas, M.; Jaffery, S.H.I.; Khan, A.; Khan, M. Development and Analysis of Tool Wear and Energy Consumption Maps for Turning of Titanium Alloy (Ti6Al4V). J. *Manuf. Process.* **2021**, *62*, 613–622.
24. Momeni, A.; Dehghani, K. Characterization of Hot Deformation Behavior of 410 Martensitic Stainless Steel Using Constitutive Equations and Processing Maps. *Mater. Sci. Eng. A* **2010**, *527*, 5467–5473.
25. Derazkola, H.A.; García Gil, E.; Murillo-Marrodán, A.; Méresse, D. Review on Dynamic Recrystallization of Martensitic Stainless Steels during Hot Deformation: Part I—Experimental Study. *Metals* **2021**, *11*, 572. [CrossRef]
26. Kondo, Y. Behaviour of Copper during High-Temperature Oxidation of Steel Containing Copper. *ISIJ Int.* **2004**, *44*, 1576–1580.
27. Yin, L.; Balaji, S.; Sridhar, S. Effects of Nickel on the Oxide/Metal Interface Morphology and Oxidation Rate During High-Temperature Oxidation of Fe–Cu–Ni Alloys. *Metall. Mater. Trans. B* **2010**, *41*, 598–611.
28. Zhou, X.; Wang, M.; Fu, Y.; Wang, Z.; Li, Y.; Yang, S.; Zhao, H.; Li, H. Effect of Borides on Hot Deformation Behavior and Microstructure Evolution of Powder Metallurgy High Borated Stainless Steel. *Mater. Charact.* **2017**, *124*, 182–191.
29. Wang, R.; Zhang, W.; Li, Y.; Li, D.; Kang, Y.; Yang, X.; Eckert, J.; Yan, Z. Stress-Strain Behavior and Microstructural Evolution of Ultra-High Carbon Fe-C-Cr-V-Mo Steel Subjected to Hot Deformation. *Mater. Charact.* **2021**, *171*, 110746.
30. Lin, Y.C.; Li, L.T.; Xia, Y.C.; Jiang, Y.Q. Hot Deformation and Processing Map of a Typical Al–Zn–Mg–Cu Alloy. *J. Alloys Compd.* **2013**, *550*, 438–445.
31. Mirzadeh, H.; Najafizadeh, A.; Moazeny, M. Flow Curve Analysis of 17-4 PH Stainless Steel under Hot Compression Test. *Metall. Mater. Trans. A* **2009**, *40*, 2950–2958.
32. Wang, M.; Wang, R.; Wang, G.; Meng, L. Hot Behavior of Supercritical Martensitic Stainless Steel at Elevated Temperature. *J. Cent. South Univ.* **2017**, *48*, 1448–1457.
33. Chegini, M.; Aboutalebi, M.R.; Seyedein, S.H.; Ebrahimi, G.R.; Jahazi, M. Study on Hot Deformation Behavior of AISI 414 Martensitic Stainless Steel Using 3D Processing Map. *J. Manuf. Process.* **2020**, *56*, 916–927.
34. Ding, S.; Khan, S.A.; Yanagimoto, J. Constitutive Descriptions and Microstructure Evolution of Extruded A5083 Aluminum Alloy during Hot Compression. *Mater. Sci. Eng. A* **2018**, *728*, 133–143.
35. Lin, Y.C.; Dong, W.Y.; Zhou, M.; Wen, D.X.; Chen, D.D. A Unified Constitutive Model Based on Dislocation Density for an Al-Zn-Mg-Cu Alloy at Time-Variant Hot Deformation Conditions. *Mater. Sci. Eng. A* **2018**, *718*, 165–172. [CrossRef]
36. Prasad, Y.V.; Gegel, H.L.; Doraivelu, S.M.; Malas, J.C.; Morgan, J.T.; Lark, K.A.; Barker, D.R. Modeling of Dynamic Material Behavior in Hot Deformation: Forging of Ti-6242. *Metall. Trans. A* **1984**, *15*, 1883–1892. [CrossRef]
37. Wu, R.; Liu, Y.; Geng, C.; Lin, Q.; Xiao, Y.; Xu, J.; Kang, W. Study on Hot Deformation Behavior and Intrinsic Workability of 6063 Aluminum Alloys Using 3D Processing Map. *J. Alloys Compd.* **2017**, *713*, 212–221. [CrossRef]

38. Xu, L.; Chen, L.; Chen, G.; Wang, M. Hot Deformation Behavior and Microstructure Analysis of 25Cr3Mo3NiNb Steel during Hot Compression Tests. *Vacuum* **2018**, *147*, 8–17.
39. Ziegler, H. An Introduction to Thermomechanics; Zürich, Switzerland, 1983. Available online: https://api.pageplace.de/preview/DT0400.9780444598936_A23553189/preview-9780444598936_A23553189.pdf (accessed on 21 March 2025).
40. de Moura, A.N.; de Alcântara, C.M.; Vieira, E.A.; da Silva Labiapari, W.; da Cunha, M.A.; de Oliveira, T.R.; Orlando, M.T. Microstructure, Crystallographic Aspects, and Mechanical Properties of AISI 420 Martensitic Stainless Steel after Different Thermomechanical Process Routes. *Mater. Chem. Phys.* **2023**, *305*, 127723.
41. Zhang, B.; Huang, D.; Zhang, Y.; Wang, X.; Wang, T. Hot Deformation Behavior and Hot Working Map of 23Cr-8Ni Stainless Steel. *Met. Form. (Hot Work.)* **2023**, *8*, 44–48. [CrossRef]

metals

MDPI

Article

Effect of B2 Precipitation on Hot Ductility of Fe–22Mn–9Al–0.6C Low-Density Steel

Jun Wang †, Tinghui Man *,†, Yihao Zhou, Xicheng Wei and Han Dong

School of Materials Science and Engineering, Shanghai University, Shanghai 200444, China; wj1415477560@shu.edu.cn (J.W.); zyh13657406066@163.com (Y.Z.); wxc1028@shu.edu.cn (X.W.); 13910077790@163.com (H.D.)

* Correspondence: mantinghuilove@163.com

† These authors contributed equally to this work.

Abstract: Fe–Mn–Al–C low-density steels are regarded as promising materials applied in the automotive industry to achieve the minimization of vehicular emissions and fuel consumption. This study investigates the high-temperature strength and hot ductility of Fe–22Mn–9Al–0.6C low-density steel through high-temperature tensile tests at 800–950 °C. The high-temperature strength decreases with an increasing deformation temperature. This indicates that the precipitation of B2 reduces the hot ductility during the hot deformation of steel, where the results are consistent with those during the solid-solution treatment at 800–950 °C with a holding time of 0.5 h. Furthermore, at 800 °C the γ transforms into a mixture of α + DO_3 and κ-carbide precipitates. A transformation of $\kappa + DO_3 \rightarrow B2$ occurs in the temperature range of 850–900 °C, and at this point the κ-carbide dissolves into the matrix and B2 is generated, resulting in a significant decrease in hot ductility. As the temperature increases up to 950 °C, B2 emerges and transforms into the δ phase, and the κ-carbide precipitates along the γ/γ grain boundaries. The precipitation of B2 during high-temperature treatments in Fe-Mn-Al-C low-density steels is the critical factor affecting hot ductility, leading to crack generation; therefore, it is extremely essential to prevent the temperature interval of B2 precipitation during hot deformation processes.

Keywords: Fe–Mn–Al–C low-density steels; hot ductility; B2 precipitate; secondary phase transformation

1. Introduction

Pressing environmental issues have led to a higher demand for lightweight automobiles [1]. Unfortunately, the traditional method of improving the strength of advanced high-strength steels cannot meet the requirements. In such cases, Fe–Mn–Al–C low-density steels are widely used due to their lower density and excellent combination of strength and ductility [2–17].

Some studies have reported the mechanical properties of Fe–Mn–Al–C low-density steels where the ultimate tensile strength, total elongation, and yield strength are in the ranges of 0.6–2.0 GPa, 30–100%, and 0.5–1.4 GPa, respectively [8,18–23]. Yoo et al. [19] reported that the results for the combination of strength and ductility exceeded 80 GPa·% in Fe–28Mn–9Al–0.8C low-density steel. In the study [8], an optimal aging at 450 °C resulted in the subject alloy possessing a high yield strength of 1383 MPa with 32.5% elongation in Fe–9Al–28Mn–1.8C low-density steel. The excellent combination of strength and ductility in Fe–Mn–Al–C low-density steels is closely related to their microstructure, especially the secondary phases, such as κ-carbide, B2, DO_3, and β-Mn, owing to the high contents of Mn, Al, and C, which provide an important strengthening effect [3,12,24–27].

κ-carbide forms during treatment at 500 °C to 900 °C in Fe–Mn–Al–C low-density steels, and fine intragranular κ-carbide is considered to play a positive role in improving the yield stress [3,6]. In addition to κ-carbide, B2 and DO_3 ordered phases can also appear in Fe–Mn–Al–C low-density steels. The B2 ordered phase exhibits various morphologies in Fe–Mn–Al–C low-density steels: (1) coarse and elongated B2 bands, (2) polygonal B2 particles formed at grain boundaries, and (3) sharp-edged B2 particles within an austenite matrix [12,25–30]. B2 precipitates from the α when the temperature reaches the ordering temperature range, and the growth of the B2 is believed to follow the classical grain growth kinetics [31]. In the Fe–Al phase diagram, the Al belongs to an α stabilizer. When the atomic fraction of Al is higher than 20%, the body-centered cubic (BCC) structure mutates from a disordered structure to an ordered structure, becoming a DO_3 phase with a stable Fe_3Al structure. The stabilized DO_3 structure is transformed into a B2-structured Fe–Al phase when the temperature is higher than 530 °C [32]. Chao et al.'s [33] study reported that DO_3 + κ-carbide appeared on the austenite grain boundaries in Fe–28.6Mn–9.8Al–1.0C–0.8Si aged at 725 °C and 750 °C for 2 h, while a B2 + DO_3 phase was detected after aging at 800 °C for 1 h. After aging at 840 °C for 1 h, the κ-carbide disappeared. The authors of [33] suggested that a disordered BCC phase can be formed within the temperature range of 800–840 °C, and that a transformation occurs between the ordered phases during quenching at 840 °C, resulting in the formation of the B2 phase. Depending on the chemical composition and heat treatment conditions, β-Mn phases can also form in the microstructure of low-density steels. β-Mn phases with lath-shaped morphology appeared in Fe–31Mn–11Al–0.9C aged at 550 °C for 5 h [34].

Despite the excellent combination of strength and ductility in Fe–Mn–Al–C low-density steels at room temperature due to the precipitations, the formation of the secondary phases along the grain boundaries can cause cracks and weaken the high-temperature performance [35–38]. Moon et al. [36,37] investigated the high-temperature tensile properties of Fe–24Mn–8Al–1C and Fe–30Mn–9Al–1C low-density steels, showing that the ductility was greatly reduced at temperatures of 600–900 °C due to the coarse κ-carbide formed in the matrix and along the austenite grain boundaries. The coarse κ-carbide strengthened the matrix but weakened the solid-solution-depleted regions adjacent to the grain boundaries, leading to high-temperature embrittlement and intergranular cracking. Lee et al. [38] reported that the strength increased dramatically yet the ductility exhibited a considerable decrease in Fe–31.4Mn–11.4Al–0.89C low-density steel, which was attributed to β-Mn precipitation at the α/γ interfaces and γ/γ grain boundaries during aging at 550 °C. The decrease in the hot ductility is of great importance in limiting the hot processing of Fe–Mn–Al–C low-density steels [39,40]. However, there has been limited research on the effects of secondary phases' precipitation on the hot ductility and deformation mechanisms of Fe–Mn–Al–C low-density steels [36–38], particularly the action of B2 and DO_3 during high-temperature deformation.

Our previous study on designed Fe–22Mn–9Al–0.6C low-density steel for automotive connecting rods found that the ends of the connecting rods exhibited cracks during the hot-forging process. To clarify the cracking of connecting rods, this study aimed to investigate the effects of secondary phases on the hot ductility of Fe–22Mn–9Al–0.6C low-density steel during high-temperature deformation. Meanwhile, the effects of the secondary phases' precipitation on the generation of cracks are discussed, with a focus on B2.

2. Material and Experimental Methods

A 200 kg ingot of Fe–22Mn–9Al–0.6C steel was prepared via induction melting in an argon atmosphere, and the chemical composition was measured by titration. The ingot was soaked completely at 1100 °C for 6 h and then hot-forged into a rod of 60 mm in diameter at the temperature range of 1000 °C to 1150 °C followed by cooling to room temperature in sand. The average density of the Fe–22Mn–9Al–0.6C steel was 6.80 g/cm^3, which was measured using a STARTORIU BP211D electronic analytical balance (STARTORIU Ltd., Gützenburg, Germany). Figure 1 shows an SEM image of the microstructure after hot-forging, showing that the microstructure of the hot-forged steel consists of austenite (γ) and ferrite (α).

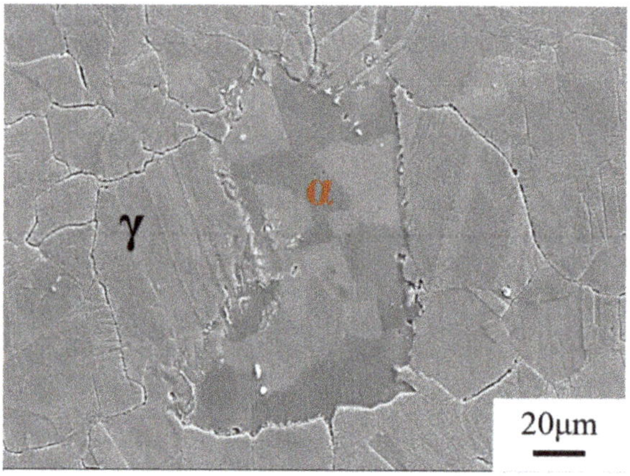

Figure 1. SEM image of the microstructure after hot-forging.

The specimens for high-temperature tension tests were cut from the hot-forged rod, machined to a diameter of 6 mm and gauge sections of 36 mm, with threads on both end sides. The size of the high-temperature tension test specimens is shown in Figure 2a. According to the standard GB/T 228-2010 [41], the high-temperature tensile tests were conducted using an INSTRON 5982 high- and low-temperature tensile testing machine (INSTRON Ltd., Canton, SD, USA) at 800 °C, 850 °C, 900 °C, and 950 °C, with a strain rate of 1.6×10^{-2} s^{-1}, and the specimens were preheated at the test temperatures for 10 min before deformation. Four tests were performed at each temperature. The specimens nearing fracture after the high-temperature tensile tests were cut to a cross-section of 10 mm × 10 mm along the longitudinal axis to explore the evolution of the secondary phases.

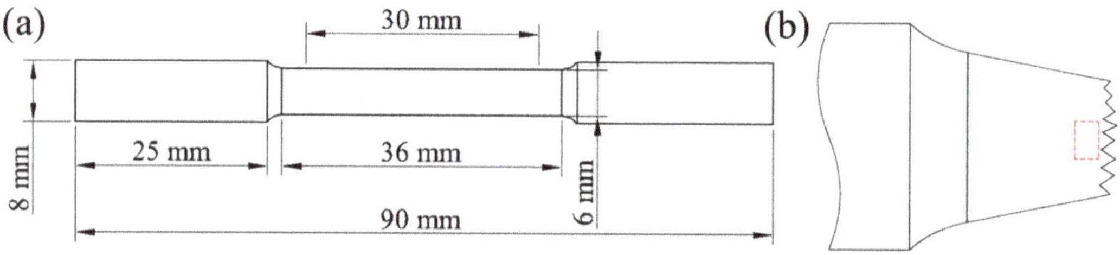

Figure 2. Schematic illustrations of (**a**) high-temperature tension test specimens and (**b**) microstructure observation in the fractured specimens (red dotted box).

Additionally, another set of specimens with dimensions of 10 mm × 10 mm × 10 mm were cut from the forged rod and solid-solution treated at 800 °C, 850 °C, 900 °C, and 950 °C for 0.5 h, before being water-quenched to room temperature to determine the secondary phases' precipitation behaviors.

The microstructure and secondary phases were identified through a combination of RIGAKU SmartLab X-ray diffraction (XRD, RIGAKU Ltd., Tokyo, Japan), ZEISS SIGMA300 scanning electron microscopy (SEM, CARL ZEISS AG Ltd., Oberkochen, Germany), and JEOL JEM-F200 transmission electron microscopy (TEM, JEOL Ltd., Tokyo, Japan). The fractured specimens and solid-solution specimens were mechanically ground and polished for the XRD measurements, and the location of the cross-section is shown in Figure 2b. The XRD measurements were performed using Cu Kα radiation with a scanning rate of 2 °/min in the scanning range of 40–100°. After the XRD measurements, the fractured specimens and solid-solution specimens were prepared for TEM and SEM observations, respectively. The thin TEM slices were cut at the fracture site of the fractured specimens to dimensions of 8 mm × 10 mm × 0.2 mm, mechanically polished to a thickness of approximately 70 μm, and then manually punched into discs with a diameter of 3 mm. The TEM discs were electrochemically etched using a LEBO TJ-100SE twin-jet electrolytic polishing machine (LEBO Ltd., Jiangsu, China) in a mixed solution of 5 vol% perchloric acid and 95 vol% methanol at −30 °C, using a voltage of 15 V. The TEM observations were performed under dark-field mode for the identification of ordered precipitates. The solid-solution specimens for the SEM observations were etched in a solution of 10% nitric acid. The SEM observations were operated at 20 KV.

3. Results

3.1. High-Temperature Tensile Properties

Figure 3 shows the high-temperature tensile properties of Fe–22Mn–9Al–0.6C low-density steel at the deformation temperatures of 800–950 °C. From Figure 3b, it can be seen that the high-temperature strength shows a decreasing trend from 170 MPa to 44 MPa as the deformation temperature increases. It is worth noting that the reduction in high-temperature strength at 800 °C to 850 °C is 92 MPa, which is much higher than the other deformation temperature spans. In addition, the elongation after fracture is approximately 60% at the deformation temperatures of 800 °C, 850 °C, and 900 °C, while for 850 °C and 900 °C it decreases to around half of the maximum (31% and 33%, respectively). It is believed that phase transformation occurs in the range of 850 °C to 900 °C, leading to the hot ductility worsening, which presumably causes and promotes crack generation in the forged connecting rods.

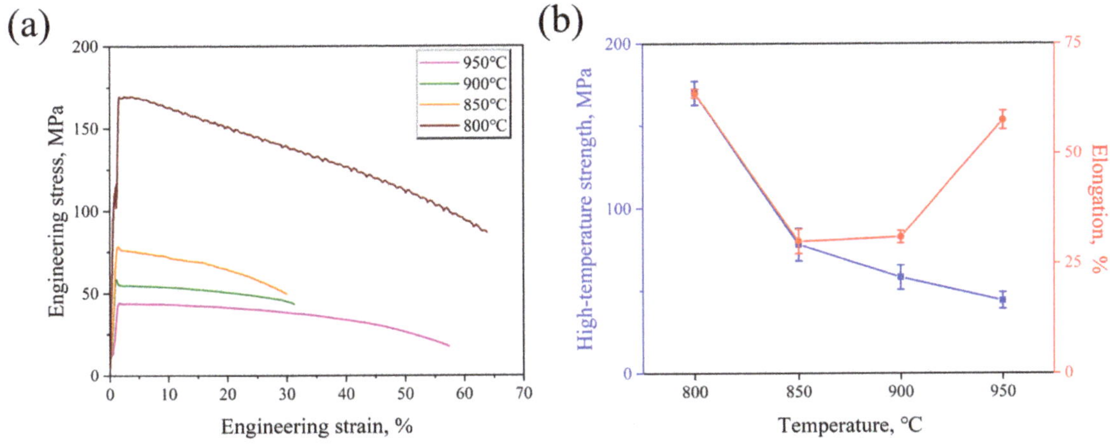

Figure 3. High-temperature tensile properties at deformation temperatures from 800 °C to 950 °C: (**a**) engineering strain–stress curves; (**b**) tensile stress and elongation.

3.2. Precipitates after High-Temperature Tensile Tests

Figure 4 shows the XRD intensity profiles of the fractured specimens subjected to high-temperature tensile tests at temperatures of 800–950 °C for Fe–22Mn–9Al–0.6C steel, where (a–d) are the diffraction peaks of the matrix, DO_3, B2, and κ-carbide, respectively. In Figure 4a, there are the diffraction peaks of face-centered cubic (FCC) and BCC structures, corresponding to γ and δ, respectively. Furthermore, it can be seen that the diffraction peaks of the DO_3(400), B2(110), and κ-carbide(111) appear at deformation temperatures of 800–900 °C, 850–900 °C, and 800 °C in Figure 4b–d, respectively.

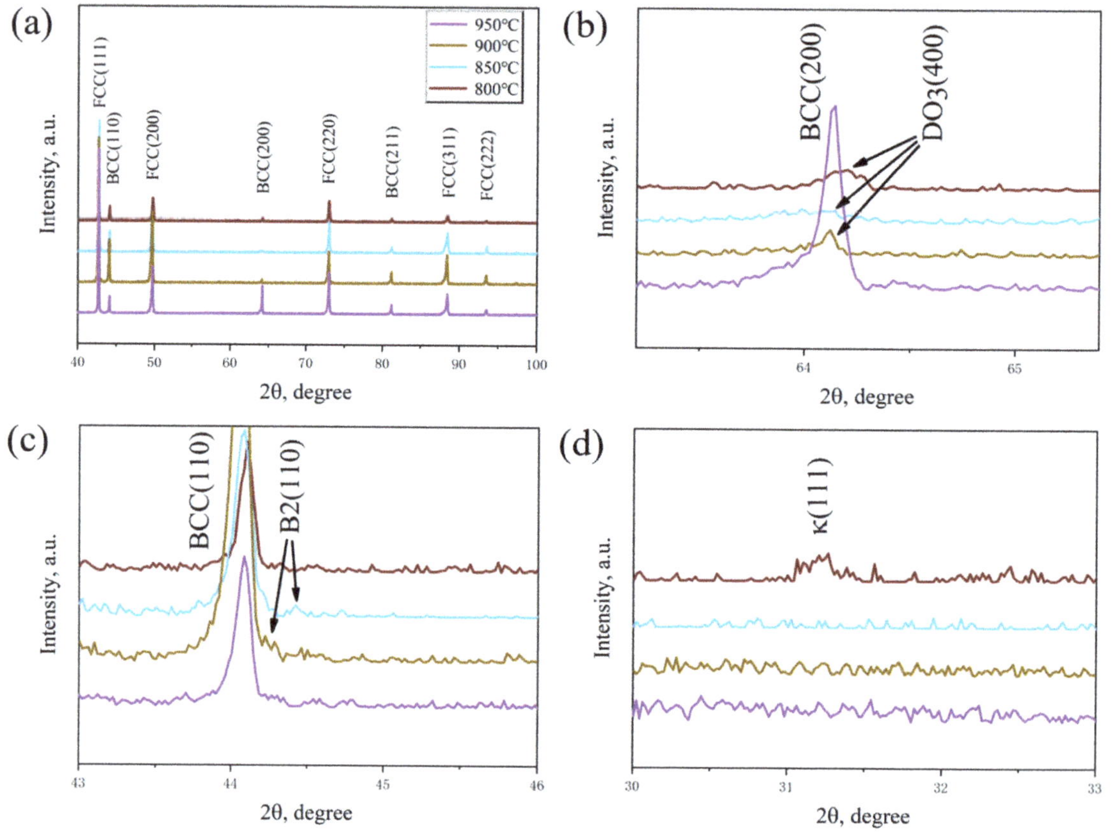

Figure 4. XRD intensity profiles of the fractured specimens at deformation temperatures of 800–950 °C: (**a**) matrix, (**b**) DO_3, (**c**) B2, and (**d**) κ-carbide.

Figure 5 shows the TEM results of the microstructure near the fracture in Fe–22Mn–9Al–0.6C after high-temperature tensile tests at 900 °C and 950 °C. The incident direction of the electron beam is [01-1]. The TEM high-resolution image and selected-area diffraction (SAD, JEOL Ltd., Tokyo, Japan) patterns of the precipitates in the fractured specimen at 900 °C are shown in Figure 5a, indicating that the diffraction pattern presents the FCC structure of γ. Moreover, weak superlattice diffraction spots were detected, which were identified as B2 and DO_3 after calibration. Figure 5b shows the TEM dark-field image and SAD patterns of the precipitate in the fractured specimen at 950 °C, revealing the presence of nanoscale precipitates within the γ grain boundaries, which were identified through calibration as κ-carbide. In the deformation temperature range of 900 °C to 950 °C, the transformation of secondary phases occurred, meaning that the B2 changed to κ-carbide. Combined with the XRD results, the precipitates were DO_3 and κ-carbide at 800 °C; as the temperature increased to 850 °C, the κ-carbide dissolved and B2 precipitated. The presence of the B2 and DO_3 remained when the deformation temperature increased to 950 °C; at higher temperatures, the DO_3 and the B2 dissolved and the κ-carbide re-precipitated.

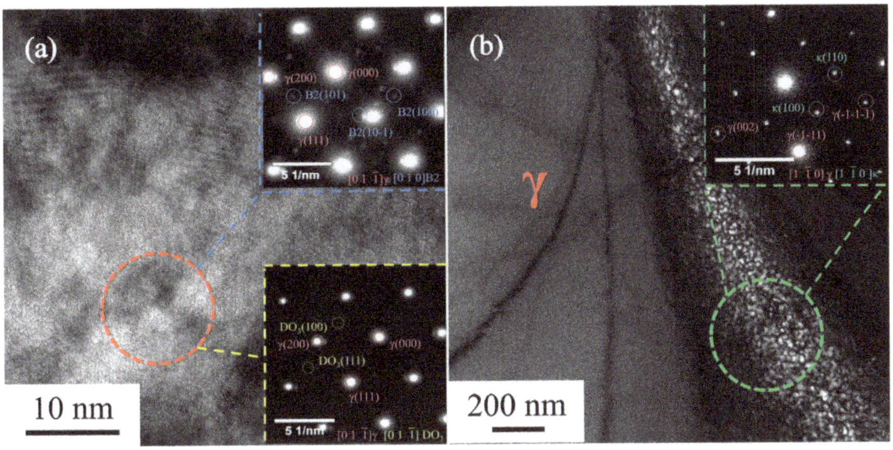

Figure 5. TEM images and SAD patterns of the precipitates in the fractured specimens at (**a**) 900 °C and (**b**) 950 °C.

3.3. Precipitates under High Temperature

Figure 6 shows the SEM images of the microstructure in the solid-solution specimens at 800–950 °C, with a holding time of 0.5 h. It can be seen that the microstructure consists of equiaxed γ and banded α (Figure 6a–c), and all grains are equiaxed when the temperature exceeds 950 °C (Figure 6d). In addition, some discrete particles (marked by red arrows) can be observed in the microstructure at 800–950 °C.

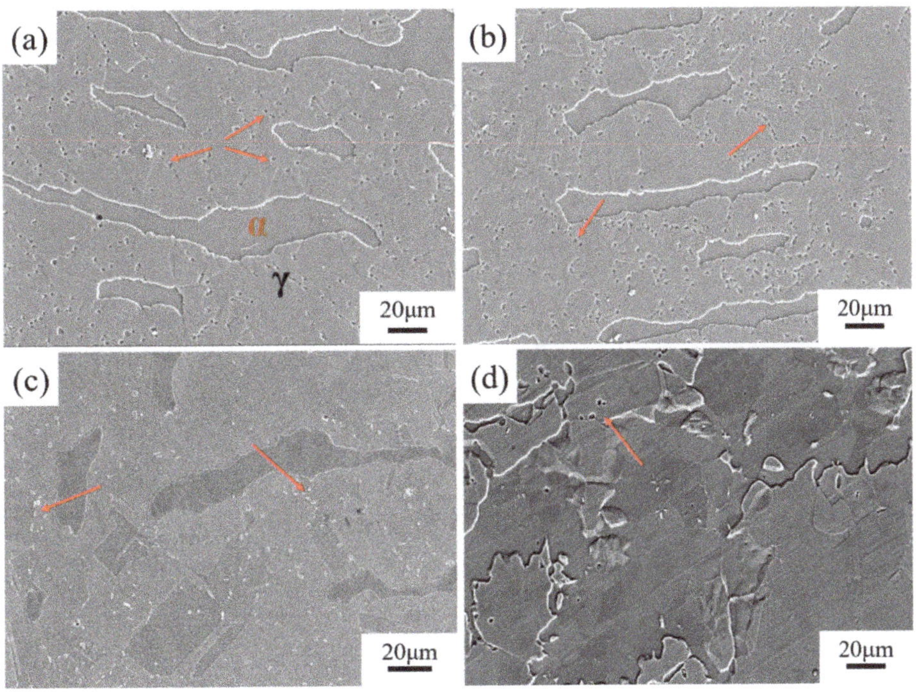

Figure 6. SEM images of the microstructure in solid-solution specimens at different temperatures: (**a**) 800 °C, (**b**) 850 °C, (**c**) 900 °C, and (**d**) 950 °C (red arrows indicate discrete particles).

Figure 7 shows the XRD intensity profiles of the solid-solution specimens subjected to temperatures of 800–950 °C with a holding time of 0.5 h for Fe–22Mn–9Al–0.6C steel, where (a–d) are the diffraction peaks of the matrix, DO_3, B2, and κ-carbide, respectively. In Figure 7a, the diffraction peaks of FCC and BCC structures can be seen, corresponding to γ and δ, respectively. In addition, it can be seen that the diffraction peaks of the DO_3(400), B2(211), and κ-carbide(111) appear at the deformation temperatures of 800–900 °C, 850–900 °C, and 800 °C/950 °C in Figure 7b–d, respectively.

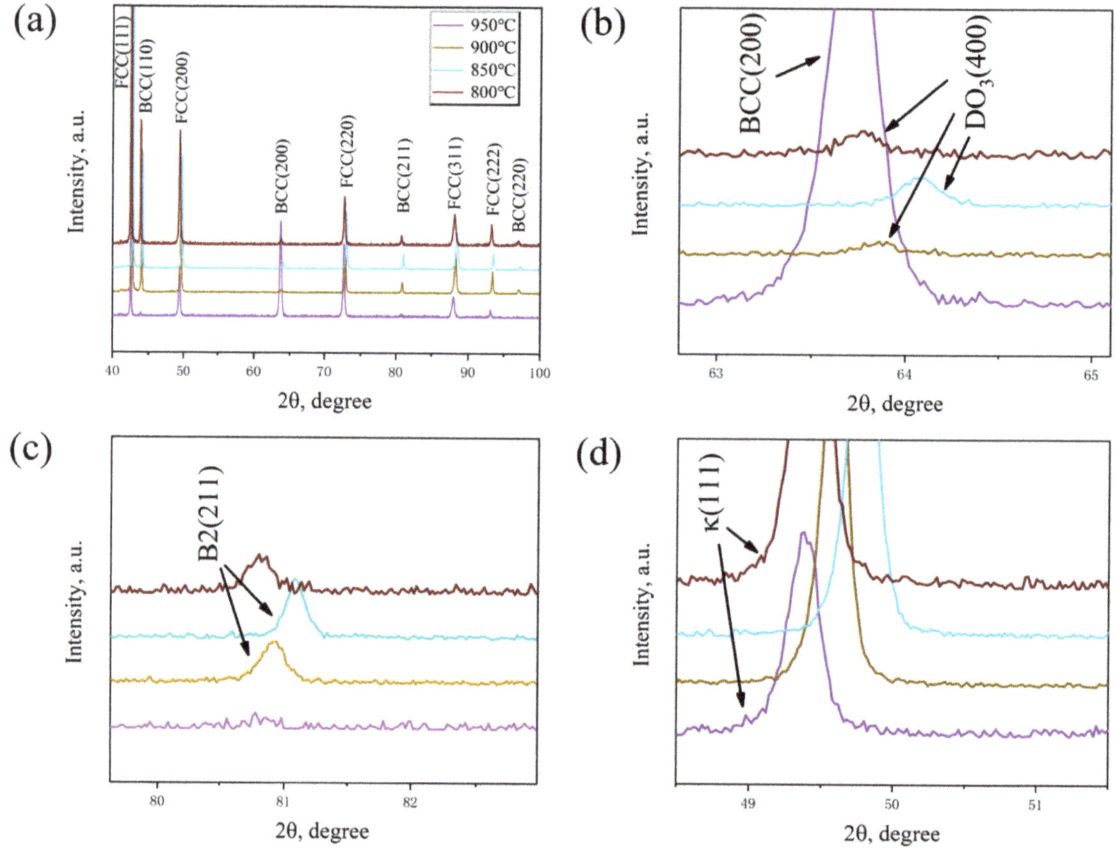

Figure 7. XRD intensity profiles of the solid-solution specimens at the temperatures of 800–950 °C: (**a**) matrix, (**b**) DO_3, (**c**) B2, and (**d**) κ-carbide.

4. Discussion

The reduction in high-temperature strength from 800 °C to 850 °C is much greater than for other deformation temperature intervals in Fe–22Mn–9Al–0.6C steel. From the XRD intensity profiles in Figure 4a–d, it can be seen that DO_3 and κ-carbide existed in addition to γ and δ at 800 °C. At 850 °C, the κ-carbide disappeared and B2 appeared. It is well known that the interactions between κ-carbide and dislocations play an important role in the strength increase of Fe–Mn–Al–C low-density steels [42]. Therefore, the sudden drop in high-temperature strength at the deformation temperature of 850 °C can be attributed to the disappearance of the κ-carbide in Fe–22Mn–9Al–0.6C steel.

In previous studies [43–45], κ-carbide was detrimental to ductility, while it increased the strength of Fe–Mn–Al–C low-density steels. In the engineering stress–strain curves of Figure 3a, the disappearance of κ-carbide does not enhance the ductility, instead causing a substantial reduction, which is well explained by the appearance of the B2 phase in Figure 4c. As the temperature increased to 850–900 °C, the κ-carbide dissolved and B2 precipitated, as shown in Figure 4c,d and Figure 7c,d. Upon further increasing the temperature to 950 °C, the B2 dissolved and the κ-carbide re-precipitated. From the TEM results in Figure 5, it can be seen that B2 and DO_3 existed at the deformation temperature of 900 °C, while κ-carbide emerged at 950 °C. The decrease in the hot ductility at the deformation temperatures of 850–900 °C is believed to be the consequence of the B2 precipitation. Therefore, it is necessary to avoid the precipitation of B2 during hot deformation processes for Fe–Mn–Al–C steel. In Figure 5b, it can be seen that the size of κ-carbide is at the nanometer scale, suggesting that the lack of κ-carbide diffraction peaks may be attributed to the small size of the κ-carbide at this temperature.

In Figure 6, some discrete particles can be observed in the solid-solution specimens at 800–950 °C. Liu et al. [46] reported that a mixture of α + DO_3 particles preferentially forms along the austenite grain boundaries during the solid-solution treatment and is retained after quenching. According to the Fe–C phase diagram, the austenite begins to undergo an ordering transformation of $\gamma \rightarrow \alpha$ when the carbon content is below 0.77% and the temperature is above Ac1, and disordered α phases form within the austenitic matrix. The formation of the α phase is closely related to the high Al content in the experimental steel, i.e., the increase in the Al content, as a ferrite stabilizer, decreases the stability of austenite under higher temperatures, thereby expanding the ferrite region and leading to the formation of a ferrite phase during the solution treatment [3]. These α phases combine with DO_3 and are retained after quenching. In Figures 4 and 7, the κ-carbide had already precipitated at 800 °C. The molecular formula for κ-carbide is $(Fe, Mn)_3(Fe, Al)C_x$ with a face-centered cubic structure. The Al and Fe (Mn) are located at the corner and face-centered positions, respectively; hence, Al–Al nearest neighbors do not exist. The DO_3-type antiphase vectors a/2<100> can produce Al–Fe nearest neighbors, which are similar to those existing in the κ-carbide structure. Therefore, κ-carbide can form not only on the austenite matrix [47], but also within the ordered DO_3 regions [48–50]. After the precipitation, the κ-carbide's rapid quenching results in smaller-sized κ-particles, which are solubilized in the austenite matrix along with α and DO_3 with a discrete particle form.

Therefore, the XRD results of both the solid-solution treatments and high-temperature tensile tests specified that two phase-transition reactions occurred at 800–950 °C. Within the temperature range of 800–850 °C, the κ-carbide formed within the α phase did not segregate internally [46] and was solidly dissolved with the discrete particles. This structure can be preserved up to 850 °C, and the $DO_3 + \kappa \rightarrow B2$ phase transition occurs in the mixtures at 850–900 °C [51]. When maintained at this temperature, a significant amount of B2 is generated, and the κ-carbide dissolves. The B2 is generated from discrete particles by ordered transformation and distributed along the austenite grain boundaries, leading to a transition in the fracture mode of the steel towards brittle fracture [27], which results in a significant reduction in the hot ductility of low-density steel. Moreover, in B2-type structures, unlike the DO_3 structure, a/4<111> displacements produce Al-Al nearest neighbors; therefore, B2 is not favorable for the formation of κ-carbide precipitates. Therefore, aside from participating in phase transformation reactions, this could be another crucial factor contributing to the disappearance of κ-carbide between 850 °C and 900 °C. As the temperature increases, the $DO_3 + \kappa \rightarrow B2$ phase transition terminates. The B2 gradually decomposes and transforms into the δ phase at high temperatures (Figures 4b and 7b) as the temperature increases up to 950 °C, being solidly dissolved in the γ matrix. The precipitation of κ-carbide is no longer suppressed, precipitating as nanoscale particles at the γ/γ grain boundaries. The whole phase transformation reaction can be summarized as $\alpha + DO_3 + \kappa \rightarrow B2 \rightarrow \delta + \kappa$ in Figure 8. Thus, the occurrence of cracks in the connecting rod blanks can be attributed to the rapid cooling at the ends during the forging process, with some regions cooling to temperatures within the range of 850–900 °C, resulting in the formation of B2, which causes brittleness, poor hot ductility, and the initiation of cracking (stress concentration) during thermomechanical processes.

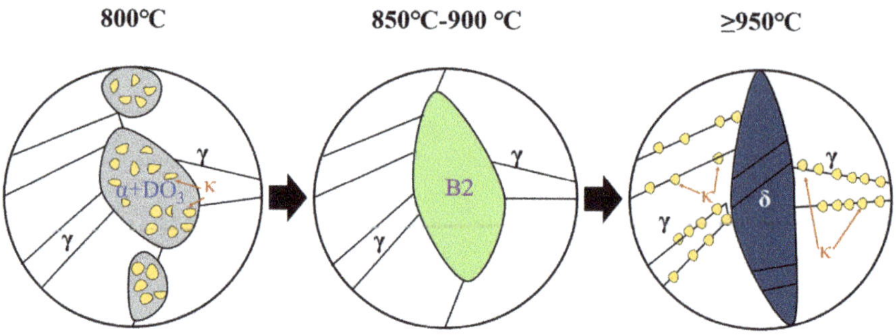

Figure 8. Schematic illustration of the secondary phases' transformations at 800–950 °C.

5. Conclusions

The hot ductility and secondary-phase transformations of Fe–22Mn–9Al–0.6C low-density steel were investigated in this study and the following conclusions are drawn:

(1) The reduction in high-temperature strength at 800 °C to 850 °C is much higher than for other deformation temperature intervals, which is presumably attributable to the dissolution of κ-carbide.

(2) The hot ductility decreases to half of the maximum, from approximately 60% to 30%, at deformation temperatures of 850–900 °C, which is presumably attributable to the formation of B2 in Fe–22Mn–9Al–0.6C low-density steel.

(3) The secondary phase transformations that occur as the temperature increases from 800 °C to 950 °C are considered to be $\alpha + DO_3 + \kappa \rightarrow B2 \rightarrow \delta + \kappa$ in the matrix.

(4) The reduction in hot ductility tends to cause cracks in the alloy during the hot-forging process and may even lead to scrapping, which needs to be taken care of during hot deformation.

Author Contributions: Conceptualization, T.M. and X.W.; methodology, T.M. and H.D.; formal analysis, J.W. and Y.Z.; investigation, J.W., Y.Z., and T.M.; resources, T.M. and H.D.; data curation, T.M. and X.W.; writing—original draft, J.W. and T.M.; writing—review and editing, J.W., T.M., and X.W.; supervision, T.M. and X.W.; project administration, T.M., X.W., and H.D.; funding acquisition, T.M. All authors have read and agreed to the published version of the manuscript.

Funding: The authors were sponsored by the Natural Science Foundation of Shanghai with grant number 23ZR1421700.

Data Availability Statement: The raw data supporting the conclusions of this article will be made available by the authors on request.

Acknowledgments: T.M. specially thanks Fort Thitipong Sangngey and Peat Wasuthorn Chaijindar for providing the most steadfast support, inspiration, and motivation that give her the strength to overcome all obstacles in her life. At the same time, T.M. thanks Ruyun Ji and Xin Wang for being with her all the way.

Conflicts of Interest: The authors declare no conflicts of interest.

References

1. Tang, D.; Mi, Z.; Chen, Y. Technology and research and development of advanced automobile steel abroad. *Iron Steel* **2005**, *40*, 1–4.
2. Churyumov, A.Y.; Kazakova, A.A.; Pozdniakov, A.V.; Churyumova, T.A.; Prosviryakov, A.S. Investigation of hot deformation behavior and microstructure evolution of lightweight Fe–35Mn–10Al–1C steel. *Metals* **2022**, *12*, 831. [CrossRef]
3. Kim, H.; Suh, D.W.; Kim, N.J. Fe-Al-Mn-C lightweight structural alloys: A review on the microstructures and mechanical properties. *Sci. Technol. Adv. Mater.* **2013**, *14*, 1–11. [CrossRef]
4. Zargaran, A.; Kim, H.S.; Kwak, J.H.; Kim, N.J. Effects of Nb and C additions on the microstructure and tensile properties of lightweight ferritic Fe–8Al–5Mn alloy. *Scr. Mater.* **2014**, *89*, 37–40. [CrossRef]
5. Sutou, Y.; Kamiya, N.; Umino, R.; Ohnuma, I.; Ishida, K. High-strength Fe–20Mn–Al–C-based alloys with low density. *ISIJ Int.* **2010**, *50*, 893–899. [CrossRef]

6. Frommeyer, G.; Brüx, U. Microstructures and mechanical properties of high-strength Fe–Mn–Al–C light-weight TRIPLEX steels. *Steel Res. Int.* **2006**, *77*, 627–633. [CrossRef]
7. Chen, S.; Rana, R.; Haldar, A.; Ray, R.K. Current state of Fe–Mn–Al–C low density steels. *Prog. Mater. Sci.* **2017**, *89*, 345–391. [CrossRef]
8. Chang, K.M.; Chao, C.G.; Liu, T.F. Excellent combination of strength and ductility in an Fe-9Al-28Mn-1.8 C alloy. *Scr. Mater.* **2010**, *63*, 162–165. [CrossRef]
9. Yao, M.J.; Welsch, E.; Ponge, D.; Haghighat, S.M.H.; Sandlöbes, S.; Choi, P.; Herbig, M.; Bleskov, I.; Hickel, T.; Lipinska-Chwalek, M.; et al. Strengthening and strain hardening mechanisms in a precipitation-hardened high-Mn lightweight steel. *Acta Mater.* **2017**, *140*, 258–273. [CrossRef]
10. Haase, C.; Zehnder, C.; Ingendahl, T.; Bikar, A.; Tang, F.; Hallstedt, B.; Hu, W.; Bleck, W.; Molodov, D.A. On the deformation behavior of κ-carbide-free and κ-carbide-containing high-Mn light-weight steel. *Acta Mater.* **2017**, *122*, 332–343. [CrossRef]
11. Hu, J.; Li, X.Y.; Meng, Q.W.; Wang, L.Y.; Li, Y.Z.; Xu, W. Tailoring retained austenite and mechanical property improvement in Al-Si-V containing medium Mn steel via direct intercritical rolling. *Mater. Sci. Eng. A* **2022**, *855*, 143904. [CrossRef]
12. Choi, K.; Seo, C.H.; Lee, H.; Kim, S.K.; Kwak, J.H.; Chin, K.G.; Park, K.T.; Kim, N.J. Effect of aging on the microstructure and deformation behavior of austenite base lightweight Fe-28Mn-9Al-0.8 C steel. *Scr. Mater.* **2010**, *63*, 1028–1031. [CrossRef]
13. Frommeyer, G.; Drewes, E.J.; Engl, B. Physical and mechanical properties of iron-aluminium- (Mn, Si) lightweight steels. *Metall. Res. Technol.* **2000**, *97*, 1245–1253. [CrossRef]
14. Imandoust, A.; Zarei-Hanzaki, A.; Sabet, M.; Abedi, H.R. An analysis of the deformation characteristics of a dual phase twinning-induced plasticity steel in warm working temperature regime. *Mater. Des.* **2012**, *40*, 556–561. [CrossRef]
15. Kim, Y.G.; Park, Y.S.; Han, J.K. Low temperature mechanical behavior of microalloyed and controlled-rolled Fe–Mn–Al–CX alloys. *Metall. Trans. A* **1985**, *16*, 1689–1693. [CrossRef]
16. Granato, A.; Hikata, A.; Lücke, K. Recovery of damping and modulus changes following plastic deformation. *Acta Metall.* **1958**, *6*, 470–480. [CrossRef]
17. Wu, Z.; Tang, Y.; Chen, W.; Lu, L.; Li, E.; Li, Z.; Ding, H. Exploring the influence of Al content on the hot deformation behavior of Fe-Mn-Al-C steels through 3D processing map. *Vacuum* **2019**, *159*, 447–455. [CrossRef]
18. Song, C.; Wang, H.; Sun, Z.; Xu, J.; Chen, H.; Yin, W. A new hot-rolled lightweight steel with ultra-high strength and good ductility designed by dislocation character and transformation strain. *Scr. Mater.* **2022**, *212*, 114583. [CrossRef]
19. Yoo, J.D.; Park, K.T. Microband-induced plasticity in a high Mn-Al-C light steel. *Mater. Sci. Eng. A* **2008**, *496*, 417–424. [CrossRef]
20. Yoo, J.D.; Hwang, S.W.; Park, K.T. Factors influencing the tensile behavior of a Fe-28Mn-9Al-0.8 C steel. *Mater. Sci. Eng. A* **2009**, *508*, 234–240. [CrossRef]
21. Ha, M.C.; Koo, J.M.; Lee, J.K.; Hwang, S.W.; Park, K.T. Tensile deformation of a low density Fe–27Mn–12Al–0.8 C duplex steel in association with ordered phases at ambient temperature. *Mater. Sci. Eng. A* **2013**, *586*, 276–283. [CrossRef]
22. Raabe, D.; Springer, H.; Gutiérrez-Urrutia, I.; Roters, F.; Bausch, M.; Seol, J.B.; Koyama, M.; Choi, P.P.; Tsuzaki, K. Alloy design, combinatorial synthesis, and microstructure–property relations for low-density Fe-Mn-Al-C austenitic steels. *Jom* **2014**, *66*, 1845–1856. [CrossRef]
23. Liu, C.; Peng, Q.; Xue, Z. Research Situation of Fe–Mn–Al–C Low-density High-strength Steel. *Mater. Rep.* **2019**, *33*, 2572–2581.
24. Lin, C.L.; Chao, C.G.; Bor, H.Y.; Liu, T.F. Relationship between microstructures and tensile properties of an Fe–30Mn–8.5 Al–2.0 C alloy. *Mater. Trans.* **2010**, *51*, 1084–1088. [CrossRef]
25. Wang, W.; Zhang, H.; Yang, M.; Jiang, P.; Yuan, F.; Wu, X. Shock and spall behaviors of a high specific strength steel: Effects of impact stress and microstructure. *J. Appl. Phys.* **2017**, *121*, 135901. [CrossRef]
26. Gutierrez-Urrutia, I.; Raabe, D. Influence of Al content and precipitation state on the mechanical behavior of austenitic high-Mn low-density steels. *Scr. Mater.* **2013**, *68*, 343–347. [CrossRef]
27. Kim, S.H.; Kim, H.; Kim, N.J. Brittle intermetallic compound makes ultrastrong low-density steel with large ductility. *Nature* **2015**, *518*, 77–79. [CrossRef]
28. Rahnama, A.; Spooner, S.; Sridhar, S. Control of intermetallic nano-particles through annealing in duplex low density steel. *Mater. Lett.* **2017**, *189*, 13–16. [CrossRef]
29. Zhang, X.; Li, J.; Wan, Y.; Wu, X.; Huang, Z. Research Progress of Ordered Precipitates in Low-density Steels. *Mater. Rep.* **2019**, *32*, 3979–3989.
30. Park, G.; Nam, C.H.; Zargaran, A.; Kim, N.J. Effect of B2 morphology on the mechanical properties of B2-strengthened lightweight steels. *Scr. Mater.* **2019**, *165*, 68–72. [CrossRef]
31. Cheng, W.C.; Cheng, C.Y.; Hsu, C.W.; Laughlin, D. Phase transformation of the L12 phase to kappa-carbide after spinodal decomposition and ordering in an Fe–C–Mn–Al austenitic steel. *Mater. Sci. Eng. A* **2015**, *642*, 128–135. [CrossRef]
32. Sundman, B.; Ohnuma, I.; Dupin, N.; Kattner, U.R.; Fries, S.G. An assessment of the entire Al-Fe system including D03 ordering. *Acta Mater.* **2009**, *57*, 2896–2908. [CrossRef]
33. Chao, C.Y.; Hwang, C.N.; Liu, T.F. Grain boundary precipitation behaviors in an Fe–9.8 Al–28.6 Mn–0.8 Si–1.0 C alloy. *Scr. Mater.* **1996**, *34*, 75–81. [CrossRef]
34. Lee, K.; Park, S.J.; Moon, J.; Kang, J.Y.; Lee, T.H.; Han, H.N. β-Mn formation and aging effect on the fracture behavior of high-Mn low-density steels. *Scr. Mater.* **2016**, *124*, 193–197. [CrossRef]

35. Savaedi, Z.; Motallebi, R.; Mirzadeh, H. A review of hot deformation behavior and constitutive models to predict flow stress of high-entropy alloys. *J. Alloys Compd.* **2022**, *903*, 163964. [CrossRef]
36. Moon, J.; Jo, H.H.; Park, S.J.; Kim, S.D.; Lee, T.H.; Lee, C.H.; Lee, M.G.; Hong, H.U.; Suh, D.W.; Raabe, D. Ti-bearing lightweight steel with large high temperature ductility via thermally stable multi-phase microstructure. *Mater. Sci. Eng. A* **2021**, *808*, 140954. [CrossRef]
37. Moon, J.; Park, S.J.; Lee, C.H.; Hong, H.U.; Lee, B.H.; Kim, S.D. Influence of microstructure evolution on hot ductility behavior of austenitic Fe-Mn-Al-C lightweight steels during hot tensile deformation. *Mater. Sci. Eng. A* **2023**, *868*, 144786. [CrossRef]
38. Lee, K.; Park, S.J.; Lee, J.; Moon, J.; Kang, J.Y.; Kim, D.I.; Suh, J.Y.; Han, H.N. Effect of aging treatment on microstructure and intrinsic mechanical behavior of Fe–31.4 Mn–11.4 Al–0.89 C lightweight steel. *J. Alloys Compd.* **2016**, *656*, 805–811. [CrossRef]
39. Renault, C.; Churyumov, A.Y.; Pozdniakov, A.V.; Churyumova, T.A. Microstructure and hot deformation behavior of FeMnAlCMo steel. *J. Mater. Res. Technol.* **2020**, *9*, 4440–4449. [CrossRef]
40. Hamada, A.S.; Karjalainen, L.P. Hot ductility behaviour of high-Mn TWIP steels. *Mater. Sci. Eng. A* **2011**, *528*, 1819–1827. [CrossRef]
41. *GB/T 228.1-202x;* Metallic Materials-Tensile Testing-Part 1: Method of Test at Room Temperature. Standardization Administration of China (SAC). State Administration for Market Regulation (SAMR): Beijing, China, 2019.
42. Kim, C.W.; Kwon, S.I.; Lee, B.H.; Moon, J.O.; Park, S.J.; Lee, J.H.; Hong, H.U. Atomistic study of nano-sized κ-carbide formation and its interaction with dislocations in a cast Si added FeMnAlC lightweight steel. *Mater. Sci. Eng. A* **2016**, *673*, 108–113. [CrossRef]
43. Brüx, U.; Frommeye, R.G.; Jimenez, J. Light-weight steels based on iron-aluminium-influence of micro alloying elements (B, Ti, Nb) on microstructures, textures and mechanical properties. *Steel Res.* **2002**, *73*, 543–548. [CrossRef]
44. Rana, R.; Liu, C.; Ray, R.K. Low-density low-carbon Fe-Al ferritic steels. *Scr. Mater.* **2013**, *68*, 354–359. [CrossRef]
45. Herrmann, J.; Inden, G.; Sauthoff, G. Deformation behaviour of iron-rich iron-aluminum alloys at low temperatures. *Acta Mater.* **2003**, *51*, 2847–2857. [CrossRef]
46. Liu, T.F.; Chou, J.S.; Wu, C.C. Effect of Si addition on the microstructure of an Fe–8.0 AI–29.0 Mn–0.90 C alloy. *Metall. Trans. A* **1990**, *21*, 1891–1899. [CrossRef]
47. Cheng, W.C. Phase transformations of an Fe–0.85 C–17.9 Mn–7.1 Al austenitic steel after quenching and annealing. *Jom* **2014**, *66*, 1809–1820. [CrossRef]
48. Choo, W.K.; Han, K.H. Phase constitution and lattice parameter relationships in rapidly solidified (Fe 0.65 Mn 0.35) 0.83 Al 0.17–xC and Fe 3 Al–xC pseudo-binary alloys. *Metall. Trans. A* **1985**, *16*, 5–10. [CrossRef]
49. Choo, W.K.; Kim, D.G. Lattice modulation and formation of lamellar duplex ferrite/cubic carbide microstructure in rapidly solidified Fe-Ni-Al-C alloys. *Metall. Mater. Trans. A* **1987**, *18*, 759–766. [CrossRef]
50. Han, K.H.; Yoon, J.C.; Choo, W.K. TEM evidence of modulated structure in Fe– Mn– Al–C austenitic alloys. *Scr. Metall.* **1986**, *20*, 33–36. [CrossRef]
51. Acselrad, O.; Kalashnikov, I.S.; Silva, E.M.; Khadyev, M.S.; Simao, R.A. Diagram of phase transformations in the austenite of hardened alloy Fe–28% Mn–8.5% Al–1% C–1.25% Si as a result of aging due to isothermal heating. *Met. Sci. Heat Treat.* **2006**, *48*, 543–553. [CrossRef]

metals

MDPI

Article

Effect of Solution Treatment on Mechanical Properties and Wear Resistance of Alloyed High-Manganese Steel

Xiya Qiao [1], Boyong Li [2,*], Xiao Han [1,*], Xiangyun Zhang [2] and Xin Yang [1]

[1] School of Materials and Metallurgy, University of Science and Technology Liaoning, Anshan 114051, China; qxy5272021@163.com (X.Q.); 18241279158@163.com (X.Y.)
[2] State Key Laboratory of Metallic Materials for Marine Equipment and Applications, Anshan 114009, China; zhangxiangyun94@163.com
* Correspondence: ansteel_lby@163.com (B.L.); tophanxiao@126.com (X.H.)

Abstract

We explored the effects of different solution treatment temperatures on the microstructure, mechanical properties and wear resistance of alloyed high-manganese steel, as well as the correlations among the three parameters. The results indicated that the average grain size of high-manganese steel first decreases and then gradually increases with an increasing solution temperature. After solution treatment, the surface hardness of the high-manganese steel is lower than that of untreated steel. At a solution treatment temperature of 1050 °C, the surface hardness value is lowest, at 253.1 HV. In addition, the optimal combination of tensile strength, yield strength and work hardening rate is observed at a solution treatment temperature of 950 °C. The friction performance test results revealed that the average friction coefficient on the surface of the high-manganese steel first decreased and then increased with the increase in solution treatment temperature. After solution treatment at 950 °C, it reached a minimum value 0.273 due to oxidation friction. Meanwhile, the hardness of the steel after friction increased rapidly because of its excellent work hardening ability, so the wear rate was relatively low, approximately 0.223×10^{-13} m^2/N, demonstrating optimal wear resistance. After solution treatment, the average grain size of high-manganese steel changed, and there was a transformation of the wear mechanism; the friction and wear mechanism shifted from a combination of particle wear and fatigue wear to adhesive wear, with particle wear as the auxiliary.

Keywords: high-manganese steel; solution treatment; solution temperature; mechanical properties; wear resistance

1. Introduction

As a traditional wear-resistant material, high-manganese steel has been widely utilized in high-wear industrial fields, including for mining equipment, railway crossings, and crusher liners [1–4]. However, evolving industrial demands have imposed increasingly complex service conditions—such as multi-axis impact loads and high-stress erosive environments—which necessitate enhanced comprehensive properties in high-manganese steel, particularly in terms of strain-hardening capacity, impact toughness, and wear resistance under extreme stress [5]. To address these challenges, researchers have employed diverse strategies to optimize the mechanical properties and wear resistance of high-manganese steel, including microalloying or re-alloying [6–9], pre-hardening treatments [10–13], and heat treatment [5,14–18].

https://doi.org/10.3390/met15090937

Feng et al. [9] investigated the microstructural evolution and impact wear behavior of Al-alloyed high-manganese steel and reported that the addition of aluminum improved the stacking fault energy of high-manganese steel, changed the strain hardening and deformation mechanism of the metal, and thus improved the wear resistance of the austenite matrix. Wang et al. [13] studied the effects of surface impact parameters on the performance of pre-hardened high-manganese steel samples. The results revealed that impact hardening increased the hardness of the top surface and within a certain depth. In addition, the tested steel exhibited good wear resistance after surface impact hardening. However, excess impact induced microcracking along the top surface, resulting in deteriorated wear resistance. Li et al. [5] adopted the progressive solution treatment method to perform heat treatment on high-manganese steel. The results showed that, compared with those produced by the traditional solution treatment method, the number of precipitates produced by the progressive solid solution decreased, the tensile strength and impact toughness of the steel significantly increased, and the wear resistance increased by nearly 23%. Zhou et al. [15] studied the influence of two novel heat-treatment processes on the enhancement of yield strength and hardness by Ti-V-Nb alloying in high-manganese austenitic steel. The results revealed that the yield strength and impact toughness of steel after the segmented heat preservation process were obviously greater than those after the continuous heating process. Gao et al. [18] compared the friction morphology under different temperatures and reported that the microstructure was optimized after heat treatment and that the degree of bonding between the two phases increased, thereby inhibiting the detachment of particulate matter and reducing the occurrence of abrasive particles. Although these previous studies to improve the mechanical properties and wear resistance of high-manganese steels provide insights into the effects of the solution temperature on the yield strength, tensile strength, and wear resistance, the influence of the solution temperature was different for different alloy systems of high-manganese steels.

In this study, the effects of solution temperatures on microstructure, mechanical properties and wear resistance of alloyed high-manganese steel were systematically investigated. Moreover, this study provides a theoretical basis for optimizing the heat treatment process parameters of high-manganese steel and new ideas for improving the performance of new high-manganese austenite steel.

2. Experimental Materials and Procedures

2.1. Experimental Materials

The alloy composition of existing high-manganese steel was optimized, and a new type of Fe-25Mn-0.37C-3.69Cr high-manganese steel was designed and prototyped, which was melted by a vacuum high-frequency induction furnace. The smelting steps are as follows: charging → vacuuming → filling with argon → smelting → heat preservation for 30 min → after complete melting → stirring → homogenizing at above 1200 °C for 2 h → pouring → air cooling. The chemical compositions of the investigated high-manganese steel are given in Table 1. The five test specimens with dimensions of 150 mm × 22 mm × 40 mm were prepared via the wire cutting mechanism. Specimens were solution-treated at 900 °C, 950 °C, 1000 °C and 1050 °C for 0.5 h, respectively, and then water-quenched. Figure 1 shows the solution treatment diagram of high-manganese steel.

Table 1. Chemical compositions of the investigated high-manganese steels.

Elements	C	Si	Mn	Cr	Ni	Cu	Mo	P	S	Al	Fe
Weight (wt%)	0.37	0.14	25.01	3.69	0.014	0.48	0.0079	0.007	0.0083	0.009	Balance

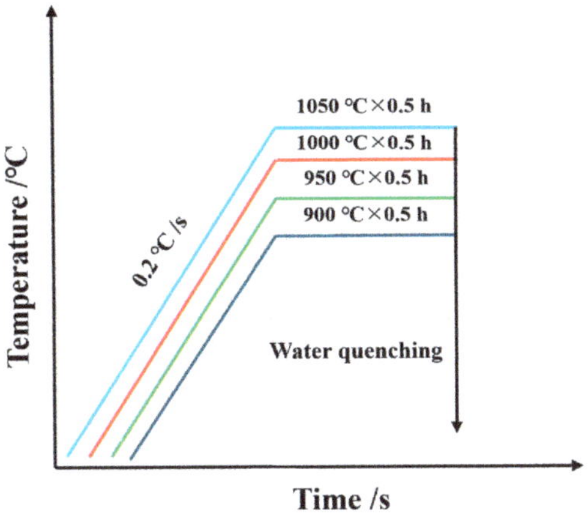

Figure 1. Solution treatment diagram of the high-manganese steel.

2.2. Mechanical Properties

The surface hardness of the steel before and after the friction test was tested via a micro-Vickers hardness meter (HV-1000, Jinan Liling Testing Machine Co., Ltd., Jinan, China). The Vickers hardness load was 9.8 N and a total of ten samples were tested. Each sample was randomly selected to measure the hardness at 5 points, and the average value was taken as the final hardness value. Tensile tests were carried out in accordance with the GB/T 228.1-2021 standard [19] on an electronic universal testing machine (CMT-5205, Shandong Shenglin Precision Machinery Equipment Co., Ltd., Jinan, China). The test temperature was 23 °C, and the tensile speed was 2 mm/min. All tensile test samples were cut along the same direction of the plate. To ensure the accuracy of the test results, each group of tests was repeated at least 3 times, and the results were averaged.

2.3. Friction Wear Test

The friction wear test was carried out using a Bruker multifunction reciprocating friction wear tester (UMT TriboLab, Bruker Corporation, Karlsruhe, Germany). A total of five samples with dimensions of 20 mm × 20 mm × 4 mm were tested. The friction pair selected for the test was a GCr15 steel ball with a diameter of 8 mm. Spherical contact friction was used, the reciprocating movement distance was 10 mm, the frequency was 200 mm/min, the load was 20 N, and the test period was 30 min.

2.4. Microstructure Observation

A total of five steel samples were mechanically polished and then electro-polished in a mixed solution of 10% perchloric acid ($HClO_4$) and 90% glacial acetic acid (CH_3COOH). The microstructures of high-manganese steel at various solution temperatures were observed using a field emission scanning electron microscope (FE-SEM, ZEISS Sigma 560, Oberkochen, Germany) equipped with an EBSD probe (Oxford Instruments, NordlysMax2, Abingdon, UK). Phase identification was performed via X-ray diffraction (XRD, D8 ADVANCE, Bruker Corporation, Karlsruhe, Germany) with a Cu-K_α target. A total of five samples were tested. The scanning rate and range of XRD analysis were 2 °/min and 20~100°, respectively. The microstructure of the wear marks was characterized by a field emission scanning electron microscope (FE-SEM). The wear marks were analyzed by an energy spectrometer (EDS). The sizes of the test samples are shown in Figure 2 (units: mm).

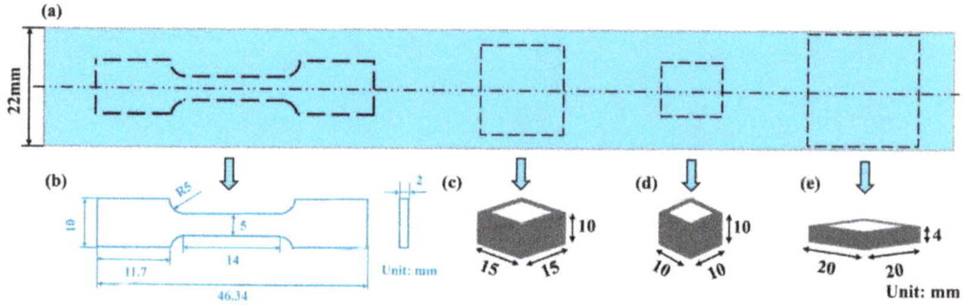

Figure 2. (**a**) Sampling locations for mechanical testing and microstructural observation; (**b–e**) dimensions of tensile samples, microstructural observation samples, surface hardness samples, and friction wear test samples.

3. Results and Discussion

3.1. Effect of Solution Treatment on Microstructures of High-Manganese Steel

The XRD patterns of the initial microstructures of the samples at different solution treatment temperatures are given in Figure 3 From the XRD results, only the characteristic diffraction peaks of austenite appeared in the specimens under different solution treatment temperatures. In addition, it was also found that the diffraction peak intensities of specimens were different. This may be related to the preferred grain orientation and texture of steels after solution treatment processes [15].

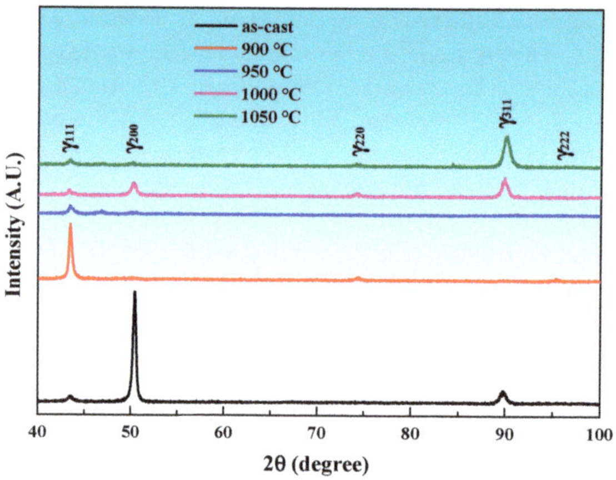

Figure 3. XRD patterns of the microstructures of high-manganese steel at various solution temperatures.

The microstructures of high-manganese steel at various solution temperatures are given in Figure 4. When the solution treatment temperature increased, part of the carbides precipitated and then dissolved, and the average grain size of high-manganese steel first decreased and then gradually increased, which was consistent with some research findings [20,21]. As shown in Figure 4a, some carbides remained in the original form at the grain boundary of the untreated steel, and the austenite grains were relatively coarse. The average grain size was approximately 100.08 μm. After 900 °C solution treatment (Figure 4b), part of the recrystallization produced equiaxed grains accompanied by residual carbides, and the average grain size was approximately 109.17 μm. As can be seen from Figure 4c, high-manganese steel undergoes optimal grain refinement at 950 °C, and the average grain size is approximately 48.21 μm. At this time, complete recrystallization produces uniform grains and annealing twins. When the solution temperature continues to rise to 1000 °C and 1050 °C, the carbides at the grain boundary dissolve into the matrix, and the austenite grain size increases significantly (Figure 4d,e). The average grain size is approximately 115.35 μm and 242.81 μm, respectively.

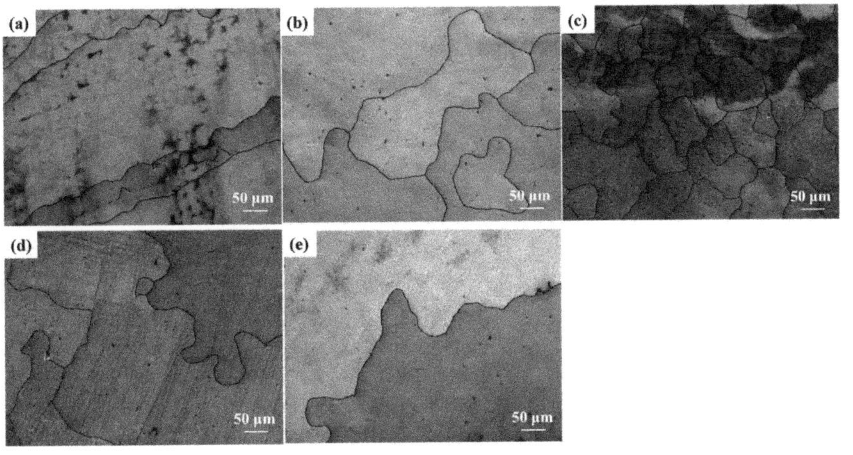

Figure 4. Microstructures of high-manganese steel at various solution temperatures: (**a**) untreated steel, (**b**) 900 °C, (**c**) 950 °C, (**d**) 1000 °C, (**e**) 1050 °C.

3.2. *Effects of Solution Treatment on the Mechanical Properties of High-Manganese Steel*

The hardness test results before and after friction for the high-manganese steel at different solution treatment temperatures are given in Figure 5a. Before the friction test, it can be observed that the hardness of the original steel sample was 291.2 HV. With the increase in the solution temperature, the hardness showed a tendency to fall. This result was consistent with the results of Xu et al.'s study [22]. With a solution temperature of 900 °C, carbides at the grain boundaries had not yet completely dissolved, which significantly improved the surface hardness of the steel. When the solution temperature increased to 1050 °C, the carbides gradually dissolved completely [23,24], resulting in the hardness of the steel sample decreasing with increasing heat treatment temperature. The steel after the 1050 °C solution treatment had the lowest hardness, approximately 251.3 HV, which can be attributed to the dissolution of carbides [22]. High-manganese steel undergoes work hardening to varying degrees after friction, resulting in a greater increase in hardness after friction than before friction. In particular, with solution temperature at 950 °C, the surface hardness is 271.1 HV to 364.2 HV, and the surface hardness of the steel is 34.3% greater than that before friction.

Tensile test results are given in Figure 5b and Table 2. A comparison of the tensile results revealed that the high-manganese steel after solution treatment had excellent yield strength and tensile strength compared to the original steel. In addition, the high-manganese steel after solution treatment further improved the elongation of the steel. The high-manganese steel has a relatively high yield/tensile strength and elongation due to the dissolution of inclusion precipitates at the grain boundaries [15]. After solution treatment at 950 °C, the steel possesses good mechanical properties, exhibiting a good combination of high yield strength (268 MPa), high tensile strength (574 MPa) and high total elongation (88.0%). Figure 5c,d showed the engineering and true stress–strain curves of high-manganese steels subjected to different solution treatment temperatures. All the samples exhibited continuous yielding behavior. The true stress–strain curve of the steel after solution treatment at 950 °C showed that the steel possesses good mechanical properties, exhibiting high true stress (1100 MPa) and large true strain (0.6).

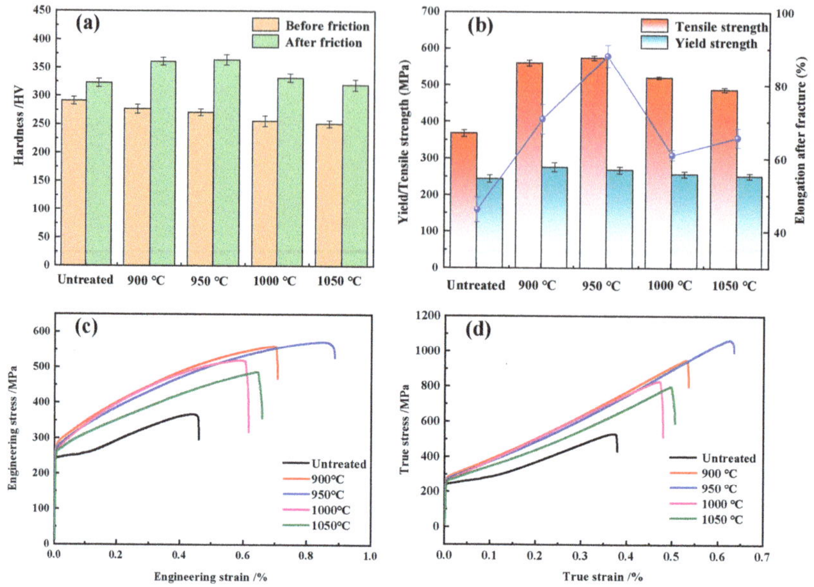

Figure 5. High-manganese steels subjected to different solution treatment temperatures: (**a**) surface hardness of high-manganese steel before and after friction, (**b**) engineering stress–strain curve, (**c**) true stress–strain curve, (**d**) tensile test results.

Table 2. Tensile test results under different heat treatment temperatures.

Materials	Yield Strength /MPa	Tensile Strength /MPa	Total Elongation /%
Untreated	244 ± 9.4	368 ± 11.2	46.0 ± 3.4
900 °C	276 ± 7.5	560 ± 13.4	70.7 ± 4.0
950 °C	268 ± 6.4	574 ± 12.0	88.0 ± 3.0
1000 °C	257 ± 3.1	520 ± 5.2	61.0 ± 1.4
1050 °C	252 ± 5.7	488 ± 7.6	65.7 ± 2.6

Figure 6 shows the room temperature tensile specimen fracture morphologies of high-manganese steels subjected to different solution treatment temperatures. As can be seen from Figure 6a, the fracture surface exhibited typical ductile fracture characteristics. The fracture surface of as-cast steel was covered with a large distribution of dimples, of varying depths and significant size differences, indicating significant plastic deformation during the stretching process [19]. The tensile fracture morphology of high-manganese steel after solution treatment also involved typical ductile fractures. In Figure 6b,c, after solution treatment at 900 °C and 950 °C, high-manganese steel had more small and deep dimples at the fracture. However, compared with 900 °C solution treatment, the size of precipitates in high-manganese steel with 950 °C solution treatment was smaller, and the dimples were more evenly distributed. With an increase in solution treatment temperatures of up to 1000 °C and 1050 °C, there were more large and shallow dimples at the fracture, and the dimples were not evenly distributed (Figure 6d,e). Some studies showed that the larger size precipitates in high-manganese steel tend to become stress concentration points during the tensile process, which accelerates the fracture of the specimen, resulting in a decrease in tensile strength and elongation of steel [19,25,26]. This is consistent with the results of tensile tests in this work.

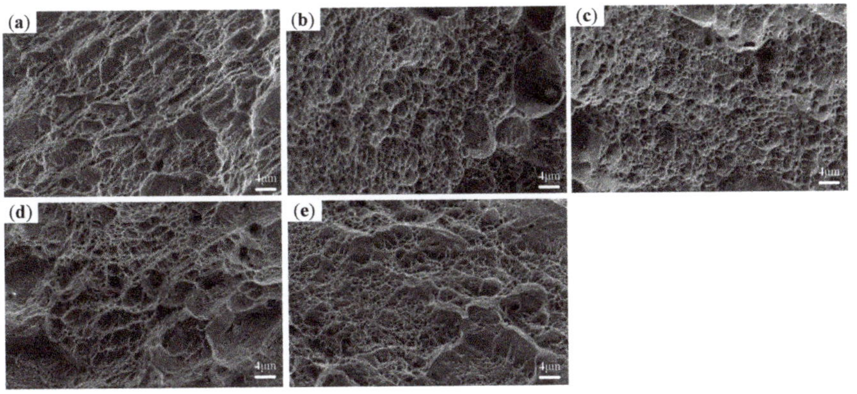

Figure 6. Room-temperature tensile fracture morphologies of high-manganese steels subjected to different solution treatment temperatures: (**a**) untreated steel, (**b**) 900 °C, (**c**) 950 °C, (**d**) 1000 °C, (**e**) 1050 °C.

Figure 7 provides the work hardening rate curves of high-manganese steels subjected to different solution treatment temperatures. The work hardening rate is generally associated with grain size, dislocation density, and deformation in high-manganese steel [27–30]. As shown in Figure 7b–f, the work hardening rate of high-manganese steel involves five stages (I, II, III, IV and V) [27,28]. The work hardening rate appears in stages II and III. During stages II and III, deformation twins continue to form, resulting in continuous work hardening. By comparing the work hardening rates at different stages, it is found that the work hardening rates at 900 °C and 950 °C are slightly higher than those of high-manganese steel at other solution treatment temperatures, indicating that the work hardening rate is higher and is most sensitive to work hardening [31].

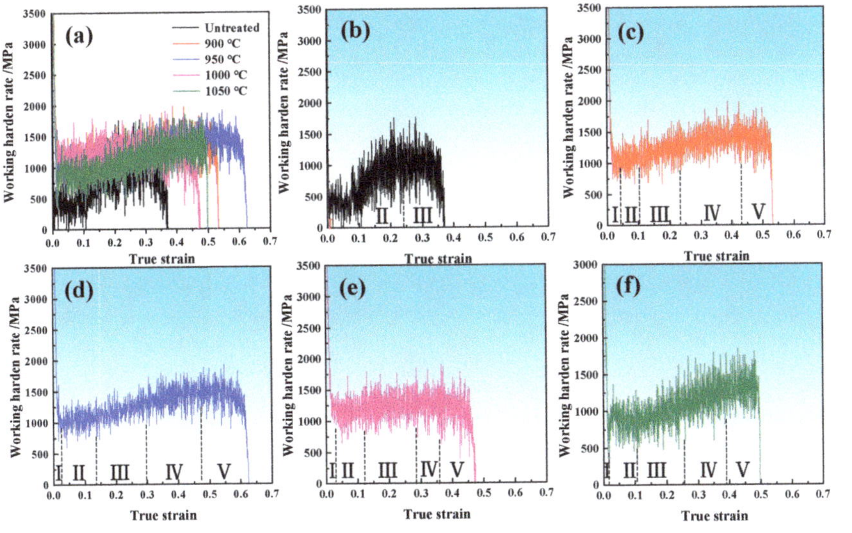

Figure 7. (**a**) Work hardening rate curves of high-manganese steels subjected to different solution treatment temperatures: (**b**) untreated, (**c**) 900 °C, (**d**) 950 °C, (**e**) 1000 °C, (**f**) 1050 °C.

3.3. *Effect of Solution Treatment on the Wear Characteristics of High-Manganese Steel*

The change curves of the friction coefficient of high-manganese steel at different solution temperatures with friction time are shown in Figure 8a. It can be seen that the friction coefficient of the high-manganese steel gradually increased and finally stabilized between 0.30 and 0.50. In general, adhesive wear happens when two friction surfaces slide or rub against each other under pressure, causing them to stick together and then tear apart; fatigue wear is a type of wear generated due to the repeated application of cyclic loads on a material [32]. At first, many particles fall off the surface due to adhesion wear, and adhesion wear gradually becomes abrasive contact wear. Moreover, due to fatigue wear, the friction contact surface forms local protrusions, leading to an increase in the roughness of the friction surface, so the friction coefficient fluctuates within a certain range [33].

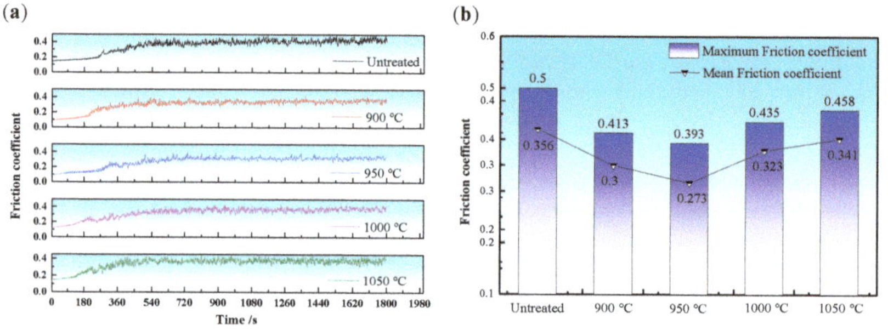

Figure 8. Different solution treatment temperatures: (**a**) friction coefficient curves and (**b**) maximum and mean friction coefficient.

Figure 8b shows the changes in the average and maximum friction coefficients with the increase in solution treatment temperature. With the increase in solution temperature, the maximum friction coefficient and the average friction coefficient first decrease and then increase. When the solution temperature was 950 °C, the mean friction coefficient reached a minimum value of 0.273. To explore the reason for the decrease in the friction coefficient at 950 °C, EDS energy spectrum analysis was performed on the wear surface, and the results are shown in Figure 9. The energy spectrum revealed that the oxidation reaction occurred during the friction process in places with deep wear marks. Owing to the combined action of friction and friction heat, each element undergoes an oxidation reaction on the friction surface, forming an oxide layer [34], increasing the surface hardness, and oxidizing the high-manganese steel, so the average friction coefficient decreases.

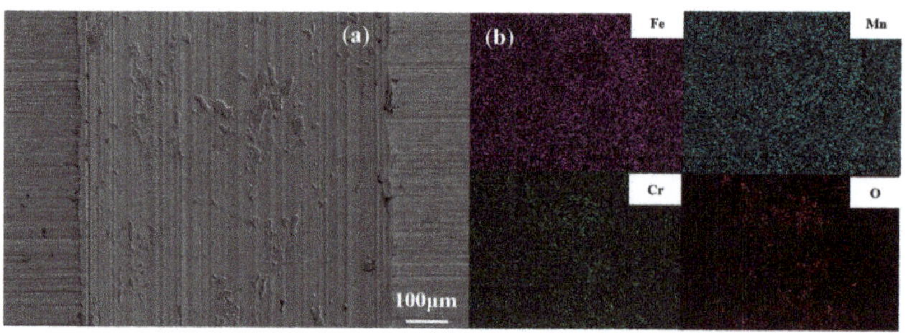

Figure 9. Wear surface morphology at 950 °C: (**a**) SEM micrograph and (**b**) EDS spectrum.

3.4. Effect of Solution Treatment on the Wear Mechanism of High-Manganese Steel

Figure 10 presents the wear surface of high-manganese steel subjected to different solution temperatures. It can be seen from Figure 10a that there was more wear debris and larger adhesive metals on the surface of the original steel wear marks, which reflects obvious abrasive wear characteristics. As shown in Figure 10b, when the solution temperature was 900 °C, the amount of surface wear debris was obviously lower, but pitting, spalling layer, and a small number of furrows remained, indicating that the wear mechanism of this high-manganese steel is mainly abrasive wear. Figure 10c shows that when the solution temperature was 950 °C, the wear surface was smoother, and small amounts of pitting, wear debris, and adhesion metals were present and evenly distributed. The wear mechanism is currently a combination of fatigue wear and adhesion wear [32,35]. When the solution temperature increased to 1000 °C, the shallow furrows and a small number of adhesion metals appeared (Figure 10d), so the wear mechanism was based on adhesive wear. When the solution temperature increased up to 1050 °C, as shown in Figure 10e, more furrows and larger adhesive metals appeared on the surface of the wear marks, the furrows were deeper, and the wear surface was rough, indicating that the degree of particle wear was very high.

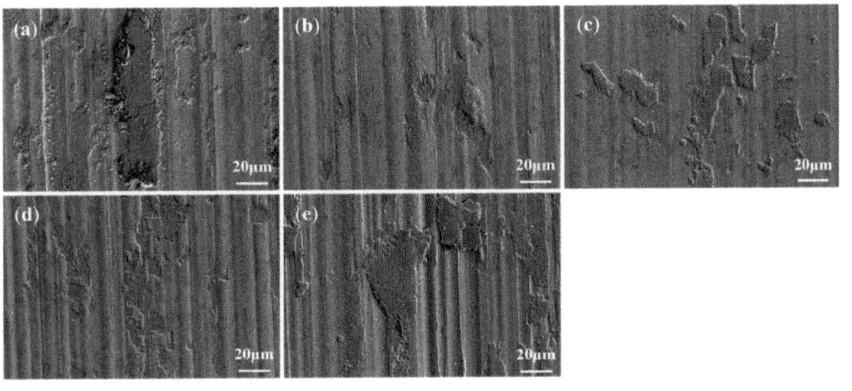

Figure 10. SEM micrographs of wear marks under different solution treatment temperatures: (**a**) untreated steel, (**b**) 900 °C, (**c**) 950 °C, (**d**) 1000 °C, (**e**) 1050 °C.

Wear resistance is typically associated with the hardness of materials. In addition, wear resistance is also closely related to the work hardening behavior [36–38]. The wear rates of the original steel and high-manganese steel at different solution temperatures are shown in Figure 11. As can be seen from Figure 11, there is a large difference between the wear rate of the original steel and the high-manganese steel at different solution temperatures, which was attributed to the work hardening effect. The surface hardness of high-manganese steel decreases after solution temperature at 950 °C, but the surface work hardening effect is improved in the friction test, the hardness is greater, and the wear resistance is excellent. The wear rate reached a minimum (0.202×10^{-13} m^2/N) when the solution temperature was 900 °C, which was consistent with the wear scar morphology results. Studies have shown that if the wear rate is faster than the strain hardening rate, a sample will deteriorate rapidly. The wear resistance of materials is closely related to the strength and hardening ability of the matrix [39]. The hardness of the high-manganese steel at a solution treatment temperature of 1000 °C and 1050 °C continued to decrease; although surface work hardening occurred during the friction process, the work hardening rate was lower than that at the solution treatment temperature of 900 °C and 950 °C, so the wear resistance was constantly reduced.

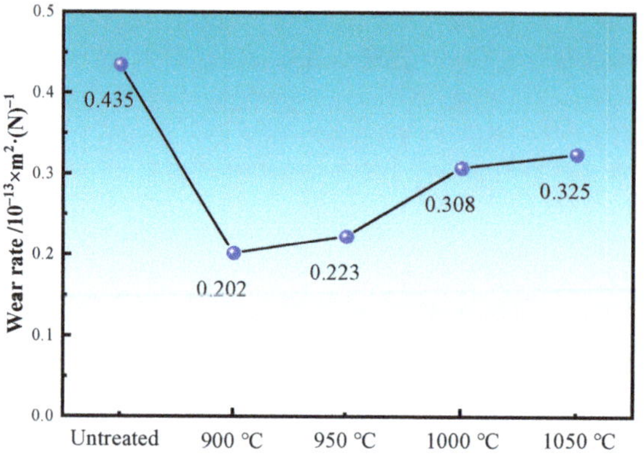

Figure 11. Wear rate under different solution treatment temperatures.

4. Conclusions

(1) With the increase in solution treatment temperature, part of the carbides precipitated and then dissolved, and the average grain size of high-manganese steel first decreased and then gradually increased. After solution treatment, the surface hardness of high-manganese steel gradually decreased. Compared with before friction, the surface hardness of high-manganese steel greatly increased after friction, especially after 950 °C solution temperature, and the surface hardness of the steel reached 364.2 HV, an increase of 34.3%. This is because high-manganese steel undergoes varying degrees of work hardening after friction. By comparing the work hardening rate of high-manganese steel, the work hardening rate is slightly higher at solution temperatures of 900 °C and 950 °C, and the hardness after friction is relatively higher.

(2) All the samples exhibited continuous yielding behavior. A comparison of the tensile results revealed that the high-manganese steel after solution treatment had excellent yield strength, tensile strength and elongation compared to the original steel. After solution temperature at 950 °C, the steel possesses good mechanical properties and wear resistance, exhibiting a good combination of high yield strength (268 MPa), high tensile strength (574 MPa) and optimal wear resistance.

(3) The friction mechanism of steel without solution treatment is mainly abrasive wear. When the solution temperature is 900–1000 °C, there is less wear debris and fewer peeling pits and grooves on the worn surface of the high-manganese steel, and the wear mechanism is mainly combined with fatigue wear and adhesive wear. When the solution temperature rises to 1050 °C, the worn surface becomes rougher, plowing grooves become more pronounced, particle wear becomes severe, and the wear resistance of the steel is reduced.

Author Contributions: X.Q.: Writing—original draft, Validation, Software, Investigation, Formal analysis, Data curation. B.L.: Writing—review and editing, Software, Formal analysis, Data curation. X.H.: Writing—review and editing, Supervision, Resources, Project administration, Methodology, Funding acquisition, Formal analysis, Conceptualization. X.Z.: Resources, Methodology, Formal analysis, Conceptualization. X.Y.: Methodology, Investigation, Data curation. All authors have read and agreed to the published version of the manuscript.

Funding: This research was funded by Joint Fund Project of State Key Laboratory of Metal Material for Marine Equipment and Application (HGSKL-USTLN (2021) 10).

Data Availability Statement: The original contributions presented in this study are included in the article. Further inquiries can be directed to the corresponding author.

Acknowledgments: The authors thank team partners from the Key Laboratory of Green Low-Carbon and Intelligent Metallurgy of Liaoning Province for their valuable contribution to this work and the preparation of this paper.

Conflicts of Interest: The authors declare that they have no known competing financial interests or personal relationships that could have appeared to influence the work reported in this paper.

References

1. Weeks, J.D. Hadfield's manganese steel. *Science* **1888**, *12*, 284–286. [CrossRef]
2. Wen, Y.H.; Peng, H.B.; Si, H.T.; Xiong, R.L.; Raabe, D. A novel high manganese austenitic steel with higher work hardening capacity and much lower impact deformation than Hadfield manganese steel. *Mater. Des.* **2014**, *55*, 798–804. [CrossRef]
3. Efstathiou, C.; Sehitoglu, H. Strain hardening and heterogeneous deformation during twinning in Hadfield steel. *Acta Mater.* **2010**, *58*, 1479–1488. [CrossRef]
4. El Fawkhry, M.K. Feasibility of new ladle-treated Hadfield steel for mining purposes. *Int. J. Miner. Metall. Mater.* **2018**, *25*, 300–309. [CrossRef]
5. Li, Z.; Wang, H.; Zhao, Y.; Wu, L.; Zhang, F.; Shan, Q. Effect of progressive solid-solution treatment on microstructures, mechanical properties and impact abrasive wear behavior of alloyed high manganese steel. *Mater. Res. Express* **2022**, *9*, 036512. [CrossRef]
6. Barbangelo, A. Influence of alloying elements and heat treatment on impact toughness of chromium steel surface deposits. *J. Mater. Sci.* **1990**, *25*, 2975–2984. [CrossRef]
7. Reyes-Calderón, F.; Mejía, I.; Boulaajaj, A.; Cabrera, J.M. Effect of microalloying elements (Nb, V and Ti) on the hot flow behavior of high-Mn austenitic twinning induced plasticity (TWIP) steel. *Mater. Sci. Eng. A* **2013**, *560*, 552–560. [CrossRef]
8. Scott, C.; Remy, B.; Collet, J.-L.; Cael, A.; Bao, C.; Danoix, F.; Malard, B.; Curfs, C. Precipitation strengthening in high manganese austenitic TWIP steels. *Int. J. Mater. Res.* **2011**, *102*, 538–549. [CrossRef]
9. Feng, X.Y.; Zhang, F.C.; Yang, Z.N.; Zhang, M. Wear behaviour of nanocrystallised Hadfield steel. *Wear* **2013**, *305*, 299–304. [CrossRef]
10. Beheshti, M.; Zabihiazadboni, M.; Ismail, M.C.; Kakooei, S.; Shahrestani, S. Investigation on simultaneous effects of shot peen and austenitizing time and temperature on grain size and microstructure of austenitic manganese steel (Hadfield). In *IOP Conference Series: Materials Science and Engineering*; IOP Publishing: Bristol, UK, 2018; Volume 328, p. 012006.
11. Meng, S.; Cui, C.Y.; Chen, K.; Zhao, K. Microstructure and mechanical properties of laser-shock-peened high-manganese steel. *Electroplat. Finish* **2020**, *39*, 760–765.
12. Hu, X.; Shen, Z.; Liu, Y.; Liu, T.; Wang, F. Influence of explosive density on mechanical properties of high manganese steel explosion hardened. *J. Appl. Phys.* **2013**, *114*, 213507. [CrossRef]
13. Wang, Z.; Yang, Y.; Chen, C.; Li, Y.; Yang, Z.; Lv, B.; Zhang, F. Effect of surface impacting parameters on wear resistance of high manganese steel. *Coatings* **2023**, *13*, 539. [CrossRef]
14. Grajcar, A.; Borek, W. Thermo-mechanical processing of high-manganese austenitic TWIP-type steels. *Arch. Civ. Mech. Eng.* **2008**, *8*, 29–38. [CrossRef]
15. Zhou, Z.; Zhang, Z.; Shan, Q.; Li, Z.; Jiang, Y.; Ge, R. Influence of heat-treatment on enhancement of yield strength and hardness by Ti-V-Nb alloying in high-manganese austenitic steel. *Metals* **2019**, *9*, 299. [CrossRef]
16. Feng, Y.; Song, R.; Peng, S.; Pei, Z.; Song, R. Microstructures and impact wear behavior of Al-alloyed high-Mn austenitic cast steel after aging treatment. *J. Mater. Eng. Perform.* **2019**, *28*, 4845–4855. [CrossRef]
17. Pu, J.; Li, Z.; Hu, Q.; Wang, Y. Effect of heat treatment on microstructure and wear resistance of high manganese steel surfacing layer. *Int. J. Mod. Phys. B* **2019**, *33*, 1940035. [CrossRef]
18. Gao, Q.; Wang, W.Z.; Yi, G.W.; Shi, P.; Fen, X.; Sun, H. Influences of annealing on microstructures and tribological properties of Ni/Ti_2AlC composites. *Tribology* **2022**, *42*, 242–253.
19. *GB/T 228.1-2021*; Metallic Materials—Tensile Testing—Part 1: Method of Test at Room Temperature. Standards Press of China: Beijing, China, 2021.
20. Yang, Y.; Liu, R.X.; Yang, H.K.; Long, J.; Han, P.; Zheng, Z. Effect of solution temperature on microstructure and properties of lightweight high manganese steel. *Heat Treat. Met.* **2023**, *48*, 113–118.
21. Sun, J.; Li, Z.; Xia, M.L.; Luo, L.; Li, J.; Huang, Z. Effect of solution treatment on microstructure and properties of Fe30Mn10Al1C low density steel. *Trans. Mater. Heat Treat.* **2024**, *45*, 121–131.
22. Xu, T.; Fu, B.; Jiang, Y.; Wang, J.; Li, G. Machine learning and experimental study on a novel Cr–Mo–V–Ti high manganese steel: Microstructure, mechanical properties and abrasive wear behavior. *J. Mater. Res. Technol.* **2024**, *31*, 1270–1281. [CrossRef]
23. Mao, Y.S.; Du, S.M.; Fu, L.H.; Zhang, Y.; Gao, Y.; Yang, J.; Bao, X. Effect of isothermal quenching process on microstructure and friction and wear properties of GCr15SiMo steel. *Tribology* **2023**, *43*, 778–790.

24. Xu, Y.R.; Zhu, T.Y.; Li, Y.J.; Wang, C.L.; Ma, T.; Hu, G.Y.; Wang, Y. Wear behavior of cobalt- based alloy brush bristles and chromium carbide coating under high line speed. *Tribology* **2022**, *42*, 1216–1225.
25. Huang, Y.; Cheng, G.G.; Li, S.J.; Dai, W.X.; Xie, Y. Effect of Ti (C, N) particle on the impact toughness of B-microalloyed steel. *Metals* **2018**, *8*, 868. [CrossRef]
26. Xu, H.F.; Zhao, J.; Cao, W.Q.; Shi, J.; Wang, C.Y.; Li, J.; Dong, H. Tempering effects on the stability of retained austenite and mechanical properties in a medium manganese steel. *ISIJ Int.* **2012**, *52*, 868–873. [CrossRef]
27. Saeed-Akbari, A.; Mosecker, L.; Schwedt, A.; Bleck, W. Characterization and prediction of flow behavior in high-manganese twinning induced plasticity steels: Part I. Mechanism maps and work-hardening behavior. *Metall. Mater. Trans. A* **2012**, *43*, 1688–1704. [CrossRef]
28. Barbier, D.; Gey, N.; Bozzolo, N.; Allain, S.; Humbert, M. EBSD for analysing the twinning microstructure in fine-grained TWIP steels and its influence on work hardening. *J. Microsc.* **2009**, *235*, 67–78. [CrossRef]
29. Renard, K.; Jacques, P.J. On the relationship between work hardening and twinning rate in TWIP steels. *Mater. Sci. Eng. A* **2012**, *542*, 8–14. [CrossRef]
30. Ding, H.; Ding, H.; Song, D.; Tang, Z.; Yang, P. Strain hardening behavior of a TRIP/TWIP steel with 18.8% Mn. *Mater. Sci. Eng. A* **2011**, *528*, 868–873. [CrossRef]
31. Lee, S.I.; Lee, S.Y.; Han, J.; Hwang, B. Deformation behavior and tensile properties of an austenitic Fe-24Mn-4Cr-0.5C high-manganese steel: Effect of grain size. *Mater. Sci. Eng. A* **2019**, *742*, 334–343. [CrossRef]
32. Yan, J.; Zhou, M.; Wu, H.; Liang, X.; Xing, Z.; Li, H.; Zhao, L.; Jiao, S.; Jiang, Z. A review of key factors affecting the wear performance of medium manganese steels. *Metals* **2023**, *13*, 1152. [CrossRef]
33. Wang, D.S.; Wang, S.Y.; Sun, S.B.; Chang, X.-T.; Yin, Y.-S. Reciprocating friction characteristics of marine Arctic steel plate at dry state room temperature. *Surf. Technol.* **2017**, *46*, 120–127.
34. Peng, Y.X.; Wang, G.F.; Zhu, Z.C.; Chang, X.D.; Lu, H.; Tang, W.; Wang, D. Friction and wear characteristics of mine hoist wire rope at low temperature. *Tribology* **2022**, *42*, 552–561.
35. Harsha, B.P.; Patnaik, A.; Banerjee, M.K.; Kozeschnik, E. Physical, mechanical, and tribological assessment of high manganese-silicon steel alloys. *Silicon* **2023**, *15*, 3305–3322. [CrossRef]
36. Yan, X.; Hu, J.; Wang, L.; Chai, Z.; Sun, W.; Xu, W. The coupled effect of thermal and mechanical stabilities of austenite on the wear resistance in a 0.2C–5Mn-1.6Si steel down to cryogenic temperatures. *Wear* **2021**, *486*, 204116. [CrossRef]
37. Rendón, J.; Olsson, M. Abrasive wear resistance of some commercial abrasion resistant steels evaluated by laboratory test methods. *Wear* **2009**, *267*, 2055–2061. [CrossRef]
38. Jost, N.; Schmidt, I. Friction-induced martensitic transformation in austenitic manganese steels. *Wear* **1986**, *111*, 377–389. [CrossRef]
39. Zhang, H.; Wang, L.; Hu, J.; Wang, G.; Xu, W. Optimum wear resistance achieved by balancing bulk hardness and work-hardening: A case study in austenitic stainless steels. *Tribol. Lett.* **2023**, *71*, 7. [CrossRef]

metals

MDPI

Article

Insight to the Microstructure Analysis of a HP Austenitic Heat-Resistant Steel Under Short-Term High-Temperature Exposure

Milica Timotijević [1,2], Olivera Erić Cekić [2,3,*], Petar Janjatović [4], Aleksandar Kremenović [5], Milena Rosić [6,*], Srecko Stopic [7] and Dragan Rajnović [4]

1 College of Applied Studies Aviation Academy, Bulevar vojvode Bojovića 2, 11158 Belgrade, Serbia; timotijevic.m@vakademija.edu.rs
2 Faculty of Mechanical and Civil Engineering in Kraljevo, University of Kragujevac, Dositejeva 19, 36000 Kraljevo, Serbia
3 Innovation Centre of the Faculty of Mechanical Engineering, University of Belgrade, Kraljice Marije 16, 11120 Belgrade, Serbia
4 Faculty of Technical Science, University of Novi Sad, Trg Dositeja Obradovića 6, 21000 Novi Sad, Serbia; janjatovic@uns.ac.rs (P.J.); draganr@uns.ac.rs (D.R.)
5 Faculty of Mining and Geology, University of Belgrade, Djusina 7, 11000 Belgrade, Serbia; aleksandar.kremenovic@rgf.bg.ac.rs
6 "Vinča" Institute of Nuclear Sciences, National Institute of the Republic of Serbia, University of Belgrade, Mike Petrovića Alasa 12-14, 11351 Belgrade, Serbia
7 Process Metallurgy and Metal Recycling, RWTH Aachen University, Intzestraße 3, 52056 Aachen, Germany; sstopic@ime-aachen.de
* Correspondence: eric.o@mfkv.kg.ac.rs (O.E.C.); mrosic@vin.bg.ac.rs (M.R.)

Abstract: The HP40Nb alloy, commonly used in the petrochemical industry as a heat-resistant material, undergoes significant microstructural changes at high temperatures. This study examined samples from the HP40Nb radiant tube used in a reformer furnace, exposed to 950, 1050, and 1150 °C for 2 and 8 h. Metallographic analysis, including optical microscopy, SEM, EDS, and XRPD, revealed that the as-cast alloy has an austenitic dendritic matrix with primary eutectic-like carbides ($M_{23}C_6$ and MC types). Prolonged exposure to high temperatures transformed the primary carbides into coarse $M_{23}C_6$ forms, losing their lamellar shape. The number of secondary carbides decreased with increasing temperature, and at 1150 °C for 480 min, secondary $Cr_{23}C_6$ carbides nearly decomposed, and Nb carbides dissolved into the austenitic matrix.

Keywords: cast heat-resistant alloy; microstructure; carbide; precipitates

1. Introduction

Centrifugally cast heat-resistant alloys of the HP (ASTM A297/ASTM A297-21a [1]) series are used for applications in the manufacture of high-temperature components in petrochemical systems, such as steam reformer tubes, radiation tubes, and ethylene cracking furnace tubes [2–5]. An HP alloy series, well known for its ASTM designation—where "H" signifies heat resistance and "P" denotes its nickel content—is primarily composed of Fe-25Cr-35Ni-0.4C (by weight) and is widely used in petrochemical furnace applications. Thanks to the unique mechanical properties under service at high temperatures, these alloys have good resistance to corrosion, high oxidation, carburization, and good creep resistance [6–10]. The reformer furnace tubes made of the HP40 (ASTM A608 [11]) alloy are designed for a nominal life of up to 100,000 h. However, the lifespan of these materials is influenced by various mechanisms, including creep, carburization, oxidation, thermal shock, and accidental overheating that occur due to the combined effects of the high service temperatures, mechanical loads, and the duration of service [12,13].

In recent years, there has been growing interest in the application of the centrifugally cast HP40Nb alloy, which has been developed by adding a small amount of niobium to the HP40 alloy [14]. It should be mentioned that the addition of niobium to the HP alloy hinders the precipitation of chromium carbides and improves its mechanical properties, increasing creep strength and creep ductility, as well as carburization resistance [12].

In the as-cast state, the microstructure of HP40Nb alloys features an austenitic matrix interspersed with intergranular and inter-dendritic eutectic-like primary Cr-rich carbides of the $M_{23}C_6$ type and Nb-rich carbides of the MC type. The microstructure of centrifugally cast HP40 reformer tubes is significantly affected by high-temperature service, leading to creep damage and the accumulation of microcracks [15]. This damage is often characterized by the presence of carbides at grain boundaries [16], coarsening of the grain boundaries due to carbide precipitation, and degradation of the mechanical properties [17,18]. These studies collectively underscore the importance of microstructural characterization in understanding the behavior of centrifugally cast steel furnace tubes, particularly in response to short or long high-temperature exposure. Thus, the formation of different carbide morphologies during processing and the expected structures in HP40Nb alloys during aging heat treatment can be understood through a combination of metallography, X-ray diffraction, and electron beam backscattered diffraction techniques [19]. Skindaras et al. [20] further explored the microstructure and mechanical properties of cast alloys with high chromium and nickel content after long-term high-temperature exposure, highlighting the emergence of carbides and other precipitates. The morphology and structure of the carbides in high-alloy steels, including HP40Nb alloys, are influenced by the aging time and secondary carbide precipitation [21]. Specifically, the eutectic carbide morphology in HP alloys changes to more complex shapes with longer aging times, and the presence of secondary carbides can lead to the coalescence of eutectic carbides [22]. The aging of HP40Nb alloys at 1000 °C under air results in the replacement of M_7C_3 carbides with $M_{23}C_6$ carbides, and the formation of a sub-surface zone depleted in chromium [23]. The inter-dendritic niobium–titanium-rich carbides in HP-NbTi alloys can have blocky or nodular morphologies [24]. The transformation of M_7C_3 to $M_{23}C_6$ and subsequent secondary carbide precipitation in austenitic matrices is a complex process influenced by various factors. The stability of the $M_{23}C_6$ phase is crucial, with its dissolution affecting the kinetics of the ferrite-to-austenite transformation [25]. The formation of $M_{23}C_6$ from M_7C_3 is influenced by temperature and time, with the former being a key factor in the transformation [26]. The presence of niobium and titanium in austenitic stainless steels can further complicate the process, with the formation of Z phase and the stability of the MX phase being important considerations [27]. The transformation of M_7C_3 to $M_{23}C_6$ and the subsequent secondary carbide precipitation are thus influenced by a range of factors, including temperature, time, and the presence of other elements in the steel.

The microstructural development of HP40 alloy micro-alloyed with Nb after overheating at a high-temperature hold is, as stated before, a complex process influenced by various factors. These studies collectively suggest that the microstructural development of an HP40 alloy with Nb is a dynamic process and further research is needed to fully understand its behavior after short-term overheating. Moreover, findings that the content of inter-dendritic carbides is a key factor in its creep performance add to the importance of a better understanding of short-term exposure [23].

Thermal instability results in accelerated microstructural aging, which unequivocally contributes to the premature and unexpected failure of tubes due to bending or cracking [28]. It is essential to recognize that there is a significant gap in our understanding of the fitness for service of tubes that endure a temperature surge without displaying any cracking or visible damage. To ensure the safe and reliable operation of tube materials throughout their lifespan, it is essential to thoroughly verify their mechanical and metallurgical properties. This careful examination acts as a safeguard, confirming that the materials possess the necessary strength and durability to withstand the demands of their application [29]. The evolution processes of carbides and other intermetallic compounds during creep exposure result in a reduction in ductility. This diminished ductility may lead to cracking due to thermal stresses occurring during shut-down and start-up operations, as well as during welding repairs [30]. Enhancing the material's performance significantly extends their lifespan, thereby reducing both capital costs and environmental impact for each unit of H_2 produced [31].

Therefore, the aim of this study was to explore the effect of short-term overheating temperatures on the microstructure and creep behavior of an HP40Nb alloy. This was achieved by discussing the microstructural evolution of HP40Nb alloys during aging, with a focus on the formation of different carbides and transition zones.

The combination of presented techniques has provided new insight into how different carbide types and morphologies form during overheating, and further highlighted the significance of microstructure control and monitoring in utilization of the reformer tubes in extreme conditions.

2. Materials and Methods

2.1. Material

The material investigated in this study was machined out of the centrifugally cast HP40Nb alloy radiant tube used in process of ethylene cracking [2,3]. The chemical composition of the tube material consists of 0.44% C, 1.11% Mn, 1.81% Si, 34.04% Ni, 26.99% Cr, and 0.63% Nb (mass %). The exact locations where the sections were cut on the reformer columns used for ethylene cracking in the respective industrial plant are not known precisely. Specimens for metallographic and mechanical tests were cut from a tube segment 350 mm long, with an outer diameter of 76.2 mm, an inner diameter of 63.5 mm, and a wall thickness of 6.35 mm, using a saw with minimal heating.

The specimens were short-term exposed at high temperatures of 950, 1050, and 1150 °C for 2 and 8 h in an electric furnace under an air atmosphere and subsequently cooled in air. The temperature was controlled by the thermocouple Pt-Pt13%Rh placed just above the samples. At least five samples were tested for each heat treatment condition.

Microstructural observations were performed on the tube samples before and after heat treatment. The morphology and chemical composition of various phases and precipitates were observed and characterized by means of metallography techniques: light microscopy (LM) and scanning electron microscopy (SEM) with secondary (SE) and backscattered electron (BSE) imaging, energy-dispersive X-ray spectroscopy (EDS), and X-ray powder diffraction (XRPD). Metallographic specimens were prepared following standard techniques, including grinding with SiC papers ranging from 180 to 2400 grit; polishing using diamond suspensions with particle sizes of 6, 3, 1, and 1/4 μm; and etching with a solution composed of 15 mL HCl, 10 mL glycerol, and 5 mL HNO_3.

2.2. Methods

The chemical composition of the service reformer tube was analyzed through the standard analytical spectrometry method, using the Type I Spark 8860 Optical Energy Spectrometer (OES) (Thermo Scientific, Waltham, MA, USA).

Metallographic specimens were prepared by the standard metallographic preparation technique: grinding (with SiC papers, grit size from 180 to 2400), polishing (diamond suspensions with 6, 3, 1, and 1/4 μm particle size), and etching with a solution of 15 mL HCl, 10 mL glycerol, and 5 mL HNO_3 [3].

The microstructure was examined using light microscopy (LM) on an Orthoplan microscope (Leitz, Oberkochen, Germany) and on a scanning electron microscope (JSM 6460 LV, JEOL, Tokyo, Japan) operating at 20 kV, and working primarily in back-scatter electron (BSE) and secondary electron (SE) imaging modes. The observed phases were analyzed using an energy dispersive X-ray analyzer system (EDS) INCA (Oxford Instruments, Abingdon, UK) in conjunction with a scanning electron microscope. It is widely recognized that EDS offers semi-quantitative chemical data and may overestimate carbon (C) content at low concentrations. However, when the carbon content in the alloy is higher, like in the case of point analysis of carbides, the margin of error is smaller. Furthermore, the size (width/depth) of carbides is larger than the interaction volume, which is around 1–2 microns for standard EDS, thus reducing any significant artefacts of the background dissolved elements. Consequently, EDS results were carefully analyzed in this study and were further supplemented by XRPD analysis to provide a more comprehensive and accurate characterization.

The Vickers hardness HV10 (ISO 6507 [32]) was determined with a test load of 98.07 N (10 kg) and a dwell time of 15 s. The testing machine was an HPO 250 (WPM, Markkleeberg, Germany). During these tests, seven measurements were taken, and the highest and lowest values were excluded to achieve a more accurate estimate of the true average.

The X-ray phase analysis (X-ray powder diffraction—XRPD) was performed on powdered samples with the help of a Smartlab powder diffractometer (Rigaku, Tokyo, Japan). Radiation from the copper anticathode $CuK_{\alpha 12}$ was used. The operating voltage on the tube was U = 40 kV, and the current was I = 30 mA. The electrolytic method of anodic extraction, successfully tested in the extraction of carbides and intermetallic phases from the matrix of different superalloys [33], was applied in this case. The electrolyte was a mixture of 94 vol% ethanol and 6 vol% HCl. The current during dissolution was 120 mA (the current density was about 0.2 A/cm^2) between the sample and titanium wire, which was also submerged in the electrolyte and served as a cathode. The sample of parallelepiped dimensions 15 × 12.5 × 6.35 mm, previously polished, was embedded in a bath of the appropriate solution and constituted the anode.

The collected fine powder after extraction was examined via Bragg–Brentano geometry (incident parallel slit, Soller slit 2.5°; incident slit, 0.5°; length limiting slit, 10 mm; receiving Slit #1, 20 mm; filter, CuK_β; receiving parallel slit, Soller slit 2.5°; receiving Slit #2, 20 mm; detector, D/teX Ultra 250 in XRF reduction mode) and in a transparent sample holder under the following experimental conditions: a range of diffraction angles of 5–130° 2θ with a step of 0.01° and a recording speed of 6°/minute. Diffraction data were processed using the software package PDXL2 (version 2.8.30, Rigaku, Tokyo, Japan). The positions of the diffraction maxima (2θ), as well as the corresponding intensities (counts), are shown graphically. On the basis of the obtained values, by comparing them with literature data and powder diffraction file (PDF) standards, the present (only) crystalline phases whose content in the sample was greater than approximately 1% were identified.

3. Results

3.1. As-Cast Microstructure

The microstructure of the as-cast alloy has a dendrite morphology with an austenitic matrix and a network of skeleton-shaped eutectic primary carbides located at the grain boundaries and between dendrites (Figure 1a). At higher magnification, according to the BSE intensity contrast, the presence of two types of second-phase particles of different morphology may be seen at the grain boundaries (Figure 1b). As can be seen in Figure 1b, one is a niobium-rich phase (white phase) having laminar-type (or skeleton form) features, whereas the other is a chromium-rich phase (dark gray phase) having fine particle-like features [18].

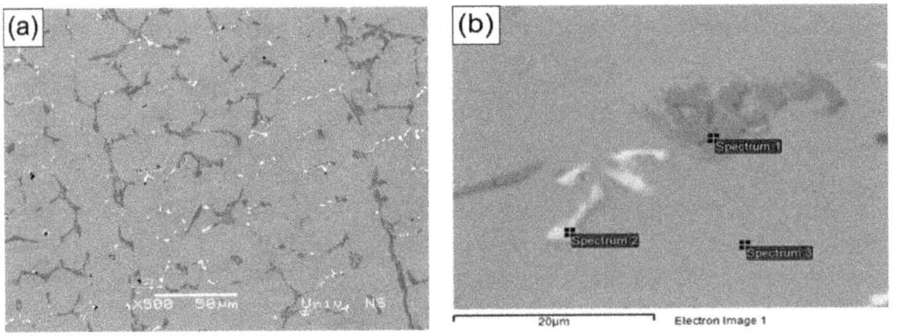

Figure 1. Microstructure of the HP40Nb alloy as cast. (**a**) SEM (BSE) microscopy at low magnification (500×); (**b**) SEM-EDS (BSE) image at high magnification (2000×).

Figure 2 shows an SEM (BSE) micrograph of the HP40Nb alloy in the as-cast condition. As seen in Figure 2, the microstructure consists of an austenitic matrix and a network of primary carbides with two types present: MC carbides (M = Nb, Ti), mainly NbC in white, and M_7C_3 carbides (M = Cr, Ni, Fe) in dark gray.

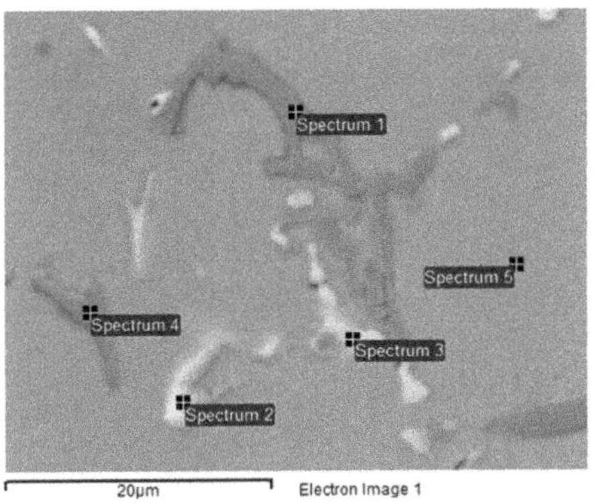

Figure 2. SEM-EDS (BSE) image of phases present in the as-cast HP40Nb alloy.

The chemical composition of the phases present in the HP40Nb alloy, as observed in Figure 2, is shown in Table 1.

Table 1. Chemical composition of the phases present in the HP40Nb alloy, corresponding to Figure 2.

Spectrum	C	Si	Ti	Cr	Mn	Fe	Ni	Nb
Spectrum 1	37.04			54.19		7.05	1.73	
Spectrum 2	43.00	0.98	0.60	10.09		9.47	8.45	27.41
Spectrum 3	43.60		3.84	5.62		3.45	2.71	40.78
Spectrum 4	17.04	2.52		44.41		19.40	16.64	
Spectrum 5	14.68	2.99		22.29	0.97	30.75	28.32	

All results are in atomic%.

Two distinct phases were identified: a niobium-rich phase, corresponding to the bright particles in Figure 2 (Spectra 2 and 3), and a chromium-rich phase, corresponding to the dark particles (Spectra 1 and 4). Furthermore, it has been noted that the two varieties of carbides exhibit different morphological characteristics. Carbides in the interdendritic boundaries appear as lamellar or skeleton forms. The M_7C_3 types show the morphological characteristics of a eutectic composite, while the MC types have a more regular morphology [33]. The niobium-rich carbides are more stable at high temperatures compared with the secondary chromium carbides [14].

3.2. XRPD Analysis

In the as-cast tube sample, the following two major crystal phases are identified: γ-Fe, whose diffraction pattern corresponds to the standard PDF # 01-081-8770, and $Cr_{23}C_6$, whose diffraction pattern corresponds to the standard PDF # 03-065-3132. However, two minor crystal phases in the amount of a few percent could be noticed: Cr_7C_3, whose diffraction pattern corresponds to the standard PDF # 01-071-3789, and Nb_3Ni_2Si, whose diffraction pattern corresponds to the standard PDF # 01-072-2171 (PDF-2 Release 2016 RDB), Figure 3. The results of the submitted chemical analysis indicate the possibility of multiple atomic replacements in the abovementioned crystal structures. The stated formulas of the crystal phases should be considered as approximate chemical formulas, i.e., $M_{23}C_6$ instead of $Cr_{23}C_6$, where M = Cr, Fe, Ni, Nb. This is in agreement with the unit cell parameters being significantly different from the standard values for γ-Fe and $Cr_{23}C_6$ (Table 2).

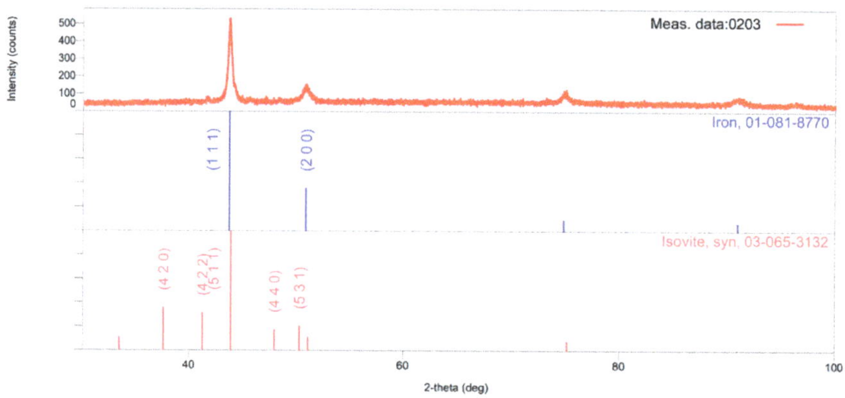

Figure 3. The X-ray powder pattern of the as-cast sample is shown in red. In the lower part of the graphic, vertical lines show the positions and intensities of the standard reflections of the abovementioned crystal phases (taken from the PDF database) and the numbers of the corresponding cards. In the first row, the positions and intensities of γ-Fe reflections are shown in blue; in the second row, the positions and intensities of $M_{23}C_6$ reflections are shown in pink.

Table 2. Unit cell parameters of significantly present crystalline phases in the as-cast sample.

Phase/Unit Cell Parameter	γ-Fe	γ-Fe PDF # 01-081-8770	$M_{23}C_6$	$Cr_{23}C_6$ PDF # 03-065-3132
a (Å)	3.58597(4)	3.5950	10.72747(15)	10.659

In the 950 °C/2 h sample, the following crystalline phases were identified: $Cr_{18.93Fe4{\cdot}07}C_6$, whose diffraction pattern corresponds to the PDF standard #01-078-1501; NbC, corresponding to the PDF standard #01-074-1222; Cr_2O_3, corresponding to the PDF standard #01-073-4336; and $Nb_3Cr_4Si_5$, corresponding to the PDF standard #01-089-2262 (PDF-2 Release 2016 RDB), as shown in Figure 4.

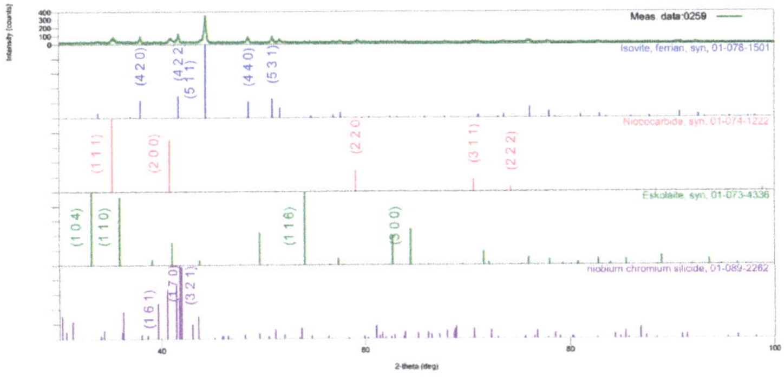

Figure 4. The XRPD pattern of the 950 °C/2 h sample is shown in red. In the lower part of the graphic, vertical lines show the positions and intensities of the standard reflections of the abovementioned crystal phases (taken from the PDF database) and the numbers of the corresponding cards.

The results of the chemical analysis provided indicate the possibility of multiple atomic substitutions within these crystalline structures. The formulas given for the crystalline phases should be considered approximate chemical formulas; for example, $M_{23}C_6$ instead of $Cr_{23}C_6$ for the carbide phase. This is in agreement with the unit cell parameters being significantly different from the standard values for NbC and $M_{23}C_6$ (Table 3).

Table 3. Unit cell parameters of significantly present crystalline phases in the 950 °C/2 h sample.

Phase/Unit Cell Parameter	NbC	$M_{23}C_6$
a (Å)	4.4062(1)	10.62020(1)

Figure 5 displays the XRPD pattern obtained from the reformer tube sample after being overheated at 950 °C for 8 h. In the reformer tube sample, the following crystal phases are identified: $Cr_{18.93}Fe_{4.07}C_6$, whose diffraction pattern corresponds to the standard PDF#, and 01-078-1501 NbC, whose diffraction pattern corresponds to the standard PDF# 01-074-1222 (PDF-2 Release 2016 RDB) (Figure 5). The results of the submitted chemical analysis indicate the possibility of multiple atomic replacements in the abovementioned crystal structures.

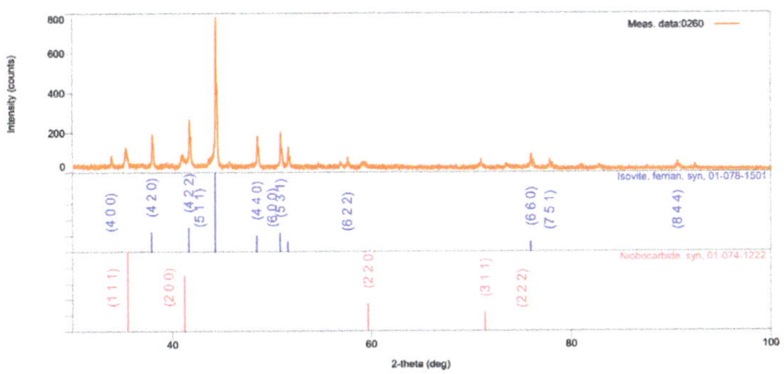

Figure 5. The XRPD pattern of the 950 °C/8 h sample is shown in orange. In the lower part of the graphic, vertical lines show the positions and intensities of the standard reflections of the abovementioned crystal phases (taken from the PDF database) and the numbers of the corresponding cards. In the first row, the positions and intensities of $Cr_{18.93}Fe_{4.07}C_6$ reflections are shown in blue; in the second row, the positions and intensities of NbC reflections are shown in pink.

The given formulas of crystalline phases should be considered as approximate chemical formulas; for example, for the carbide $M_{23}C_6$ instead of $Cr_{23}C_6$.

The unit cell parameters a (Å) of the notable crystalline phases present in the investigated 950 °C/8 h samples are displayed in Table 4.

Figure 6 shows the XRPD pattern of the reformer tube sample after overheating at 1050 °C for 2 h. The identified crystal phases include $Cr_{18.93}Fe_{4.07}C_6$, with a diffraction pattern matching the standard PDF#01-078-1501, and $Nb_{0.9}Ti_{1.1}C_2$, corresponding to the standard PDF#03-065-7915.

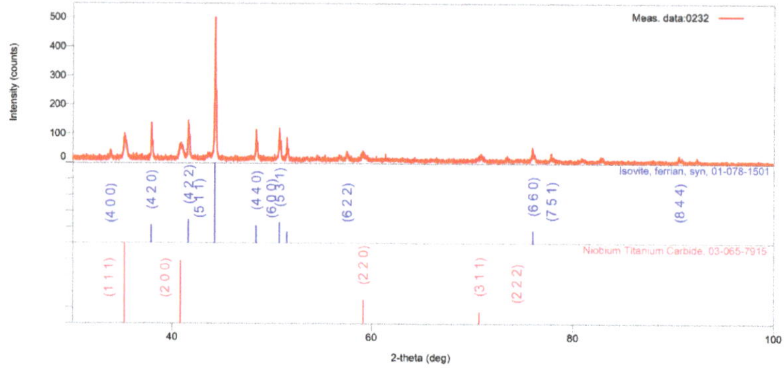

Figure 6. The X-ray powder pattern of the 1050 °C/2 h sample is shown in red. In the lower part of the graphic, vertical lines show the positions and intensities of the standard reflections of the abovementioned crystal phases (taken from the PDF database) and the numbers of the corresponding cards. In the first row, the positions and intensities of $Cr_{18.93}Fe_{4.07}C_6$ reflections are shown in blue; in the second row, the positions and intensities of $Nb_{0.9}Ti_{1.1}C_2$ reflections are shown in pink.

The results of the submitted chemical analysis indicate the possibility of multiple atomic replacements in the abovementioned crystal structures.

Figure 7 presents the XRPD pattern of the reformer tube sample after overheating at 1050 °C for 8 h. The identified crystal phases in the sample include $Cr_{18.93}Fe_{4.07}C_6$, with a diffraction pattern matching the standard PDF#01-078-1501, and $Nb_{0.9}Ti_{1.1}C_2$, corresponding to the standard PDF#03-065-7915.

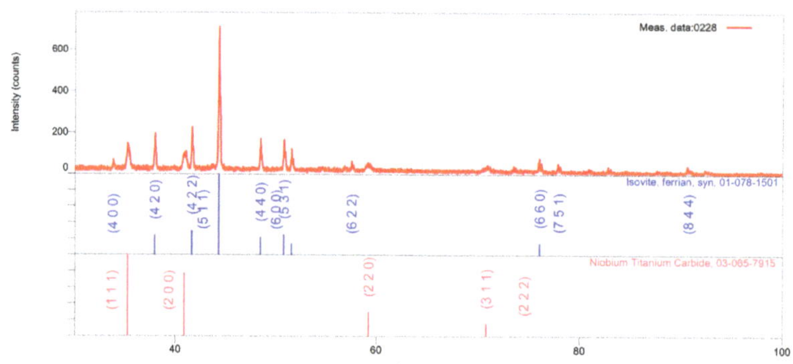

Figure 7. The X-ray powder diffraction pattern of the sample heat-treated at 1050 °C for 8 h is shown in red. Standard reflection positions and intensities for the identified crystalline phases, obtained from the PDF database, are indicated by vertical lines in the lower part of the figure along with their corresponding card numbers. Reflections for $Cr_{18.93}Fe_{4.07}C_6$ are displayed in blue (top row), while those for $Nb_{0.9}Ti_{1.1}C_2$ are shown in pink (bottom row).

Table 4. Unit cell parameters of significantly present crystalline phases in the 950 °C/8 h sample.

Phase/Unit Cell Parameter	$M_{23}C_6$
a (Å)	10.618440(5)

The submitted chemical analysis results indicate the possibility of multiple atomic replacements in the abovementioned crystal structures.

Figure 7 presents the XRPD pattern of the reformer tube sample after overheating at 1050 °C for 8 h. The identified crystal phases in the sample include $Cr_{18.93}Fe_{4.07}C_6$, with a diffraction pattern matching the standard PDF#01-078-1501, and $Nb_{0.9}Ti_{1.1}C_2$, corresponding to the standard PDF#03-065-7915.

The X-ray powder patterns of extracted powder of samples of the HP40Nb alloy after aging at 1150 °C for 2 and 8 h are shown in Figures 8 and 9, respectively.

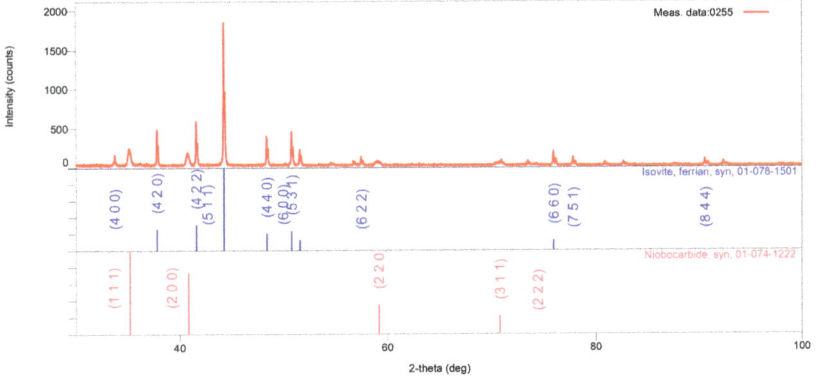

Figure 8. The X-ray powder pattern of the 1150 °C/2 h sample is shown in red. In the lower part of the graphic, vertical lines show the positions and intensities of the standard reflections of the abovementioned crystal phases (taken from the PDF database) and the numbers of the corresponding cards. In the first row, the positions and intensities of $Cr_{18.93}Fe_{4.07}C_6$ reflections are shown in blue; in the second row, the positions and intensities of NbC reflections are shown in pink.

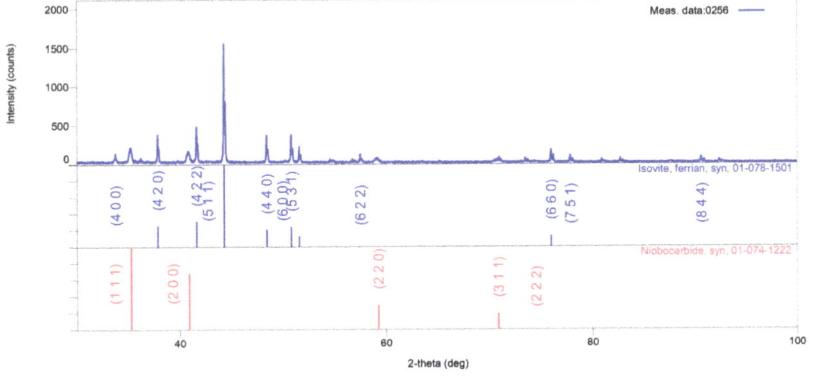

Figure 9. The X-ray powder pattern of the 1150 °C/8 h sample is shown in blue. In the lower part of the graphic, vertical lines show the positions and intensities of the standard reflections of the abovementioned crystal phases (taken from the PDF database) and the numbers of the corresponding cards. In the first row, the positions and intensities of $Cr_{18.93}Fe_{4.07}C_6$ reflections are shown in blue; in the second row, the positions and intensities of NbC reflections are shown in pink.

In case of overheating for 2 h (1150 °C/2 h), the identified phases correspond to the complex $Cr_{18.93}Fe_{4.07}C_6$ carbide of the $M_{23}C_6$-type (blue line) and to the NbC carbide (pink line), while after exposure for 8 h (1150 °C/8 h), the same phases ($Cr_{18.93}Fe_{4.07}C_6$ and NbC) were also identified.

3.3. *Microstructure After 2 h of Overheating Treatment*

Figure 10 presents representative optical micrographs (LM) illustrating the microstructure of the sample after being overheated at 950 °C, 1050 °C, and 1150 °C for 2 h.

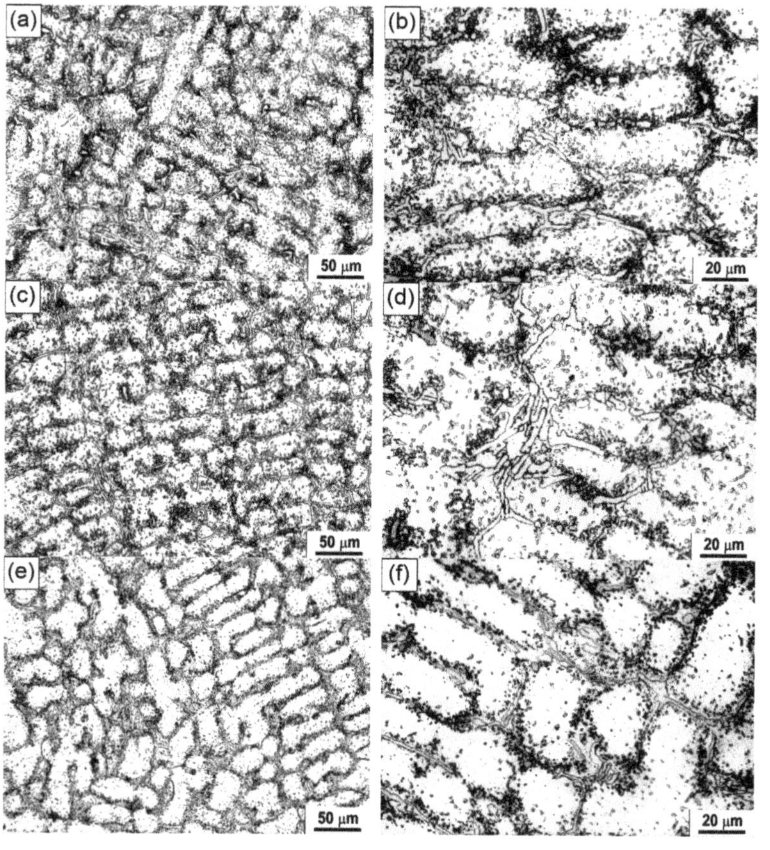

Figure 10. Light micrographs of the HP40Nb alloy after being overheated at (**a**,**b**) 950 °C, (**c**,**d**) 1050 °C and (**e**,**f**) 1150 °C for 2 h.

As illustrated in Figure 10a,b, the microstructure has experienced considerable changes following overheating at a temperature of 950 °C for 2 h. Its microstructure comprises an austenitic dendritic matrix with a discontinuous network of skeletal-like primary carbides along the grain boundaries and secondary precipitates within the grains. Many secondary carbides were precipitated in the austenitic matrix near the primary carbides and the inter-dendritic region.

Figure 11 illustrates that the microstructure experienced significant changes after overheating at 950 °C, while Table 5 presents the EDS analysis results of the phases in the sample treated at 950 °C for 2 h, as observed in the SEM micrograph in Figure 11.

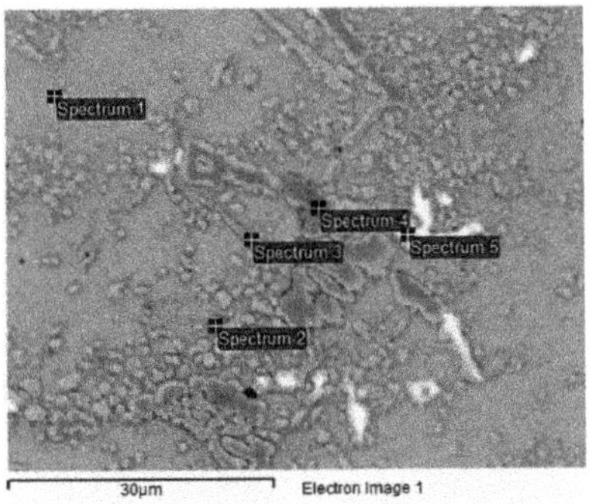

Figure 11. SEM micrograph of the HP40Nb alloy after being overheated at 950 °C/2 h.

Table 5. Chemical composition of the phases present in the HP40Nb alloy, corresponding to Figure 11.

Spectrum	C	Si	Ti	Cr	Mn	Fe	Ni	Nb	V
Spectrum 1	30.96	2.04		16.93	1.00	25.62	23.46		
Spectrum 2	38.71	2.24		20.75	0.86	18.29	19.15		
Spectrum 3	58.54	1.08		13.85	0.74	13.13	12.66		
Spectrum 4	61.09			32.49	0.55	4.33	1.33		
Spectrum 5	79.87		0.45	3.15		1.24	1.22	13.82	0.26

All results in atomic%.

The microstructure consists of an austenitic matrix (Spectra 1, 2, and 3) and two types of carbides: niobium-rich carbides (white phase, Spectrum 5) and chromium-rich carbides (dark gray phase, Spectrum 4). On the basis of Figure 10a,b and Figure 11, as well as Table 5, the dark carbides are identified as the $M_{23}C_6$ type (M = Cr) and the light carbides as the MC type (M = Nb), with traces of niobium–chromium–nickel silicide ($Nb_3Ni_2Cr_2Si_6$) and chromium oxide (Cr_2O_3) also present.

Figure 10c,d show the microstructure of the sample after overheating at 1050 °C for 2 h. The microstructure consists of an austenitic dendritic matrix and a discontinuous network of primary skeletal-shaped carbides located along the dendrite boundaries. Table 6 provides the EDS analysis results of the phases in the sample treated at 1050 °C for 2 h, as observed in the SEM micrograph in Figure 12.

Table 6. Chemical composition of the phases present in the HP40Nb alloy, corresponding to Figure 12.

Spectrum	C	Si	Ti	Cr	Mn	Fe	Ni	Nb	V
Spectrum 1	17.58	3.15		19.30	1.11	29.98	28.71	0.16	
Spectrum 2	25.06	2.40		26.53	1.32	23.00	21.68		
Spectrum 3	18.21	2.81		23.22	1.25	28.01	26.51		
Spectrum 4	25.63			60.91	1.23	8.63	3.25		0.36
Spectrum 5	56.24		0.48	6.02		1.75	2.61	32.54	0.37

All results are in atomic%.

The microstructure consists of an austenitic matrix (Spec. 1) and two types of precipitates: one rich in Nb (white phase, Spec. 6) and one rich in Cr (dark gray phase, Spec. 3, 4, and 5).

A large amount of secondary carbides has precipitated in the austenitic matrix near the primary carbides and the zones around the dendrite boundaries. According to Figures 10c,d, and 12 and Table 6, it can be observed that the primary chromium carbides completely transformed from M_7C_3 to $M_{23}C_6$. Secondary carbides are detected within the austenitic matrix near the primary carbides. It is important to note that a small quantity of acicular carbides $M_{23}C_6$ (M = Cr, Ni, Fe) was formed in the austenite matrix.

Additionally, the amount of precipitated secondary carbides increased with the increase in the superheating temperatures up to 1050 °C.

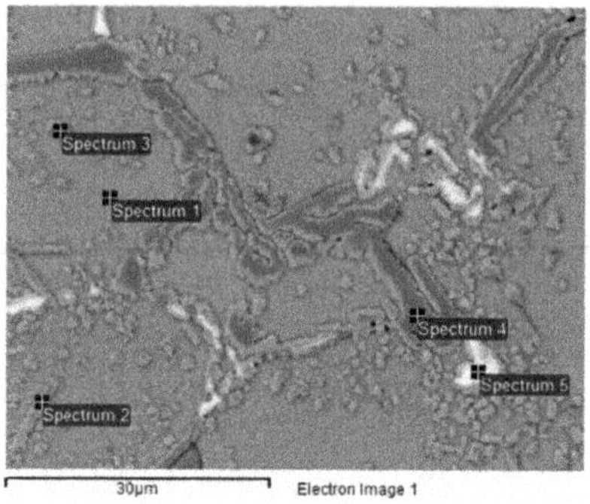

Figure 12. SEM micrograph of the HP40Nb alloy after being overheated at 1050 °C/2 h.

Figure 10e,f show the microstructure of the sample after overheating at 1150 °C for 2 h. Following this thermal treatment, the quantity of secondary carbides in the austenitic matrix and around the primary carbides was significantly reduced. The microstructure consists of an austenitic dendritic matrix and a discontinuous network of primary skeletal-shaped carbides located along the dendrite boundaries. The primary carbides have coagulated and merged. A significant amount of fine secondary carbides is present in the austenitic matrix near the primary carbides, while carbide-free zones have appeared in the inter-dendritic regions. Although the quantity of secondary carbides has decreased, their coarsening is evident [34]. Table 7 provides the EDS analysis results of the phases in the sample treated at 1150 °C for 2 h, as observed in the SEM micrograph in Figure 13. As a result of overheating at 1150 °C for 2 h, globular secondary carbides have formed in the inter-dendritic zones. On the basis of Figures 12 and 13, it can be concluded that increasing the temperature from 1050 °C to 1150 °C did not affect the type of carbides present in the microstructure.

Table 7. Chemical composition of the phases present in the HP40Nb alloy, corresponding to Figure 13.

Spectrum	C	Si	Ti	Cr	Mn	Fe	Ni	Nb	V
Spectrum 1	18.18	3.24		19.63	1.27	29.48	28.20		
Spectrum 2	34.04			47.88	0.99	10.95	6.13		
Spectrum 3	33.77			47.95	1.06	11.29	5.92		
Spectrum 4	69.05		1.77	2.9		2.54	2.41	21.12	0.21
Spectrum 5			6.65	9.83		4.17	3.18	75.33	0.58
Spectrum 6	15.89	2.24		30.73	1.23	26.11	23.80		

All results are in atomic%.

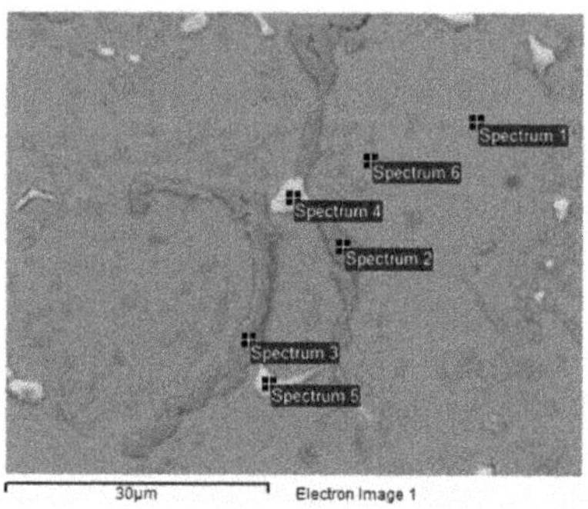

Figure 13. SEM micrograph of the HP40Nb alloy after being overheated at 1150 °C/2 h.

3.4. Microstructure After 8 h of Overheating Treatment

Figure 14a–c illustrate the microstructure of the material obtained after overheating at 950 °C, 1050 °C, and 1150 °C for 8 h.

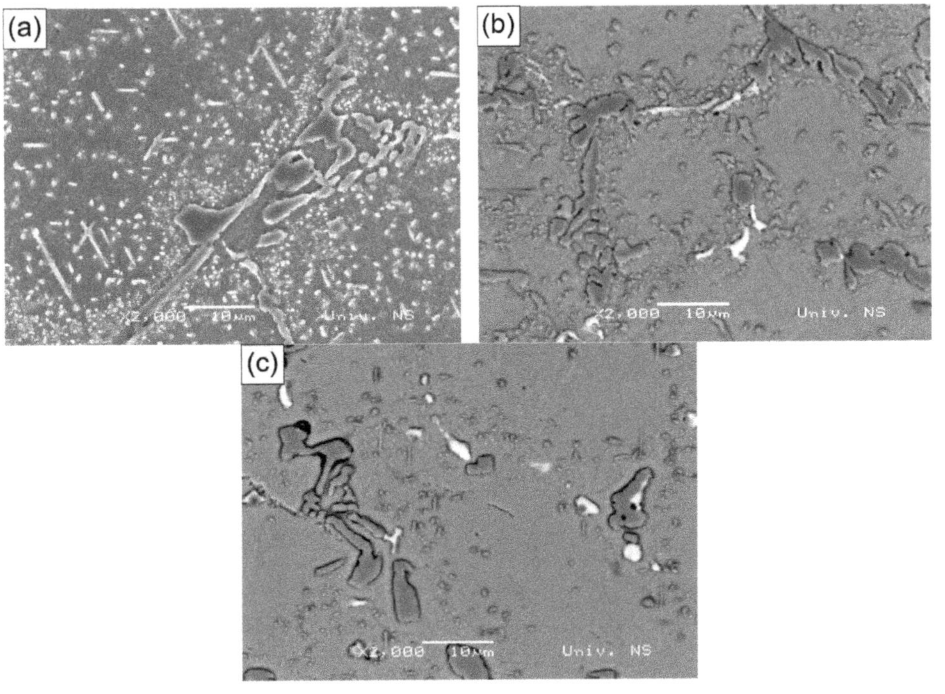

Figure 14. SEM micrographs of the HP40Nb alloy after being overheated at (**a**) 950 °C (SE), (**b**) 1050 °C (BSE) and (**c**) 1150 °C (BSE) for 8 h.

Figure 14a shows the microstructure of the HP40Nb alloy after overheating at 950 °C for 8 h. The microstructure consists of an austenitic dendritic matrix and a discontinuous network of primary skeletal-shaped carbides located along the dendrite boundaries. A large amount of secondary carbides has precipitated within the dendrites.

Figure 14b shows the microstructure of the HP40Nb alloy after overheating at 1050 °C for 8 h. A large amount of fine secondary carbides has precipitated near the primary carbides located along the dendrite boundaries. Compared with the as-cast microstructure, a finer grain structure is observed, with carbides dissolved along the austenitic grain boundaries. Unlike the as-cast structure on the boundary of primary austenitic grains, a large number of tiny carbides are present, which may be the result of the dissolution of massive carbides on the boundary and carbide islands inside the grain, as indicated in some data in the literature [4,34].

Figure 14c shows the microstructure of the HP40Nb alloy after overheating at 1150 °C for 8 h. A small amount of secondary carbides can be observed in the austenitic matrix. For samples overheated for 8 h, the increase in overheating temperature led to the dissolution of secondary carbides in the inter-dendritic region, followed by the dissolution of carbides near the primary carbides along the dendrite boundaries. On the basis of Figures 15–17, it can be observed that the primary chromium carbides (M_7C_3) transformed into $M_{23}C_6$-type carbides [35–41]. Secondary carbides were detected within the austenitic matrix near the primary carbides (Figure 15).

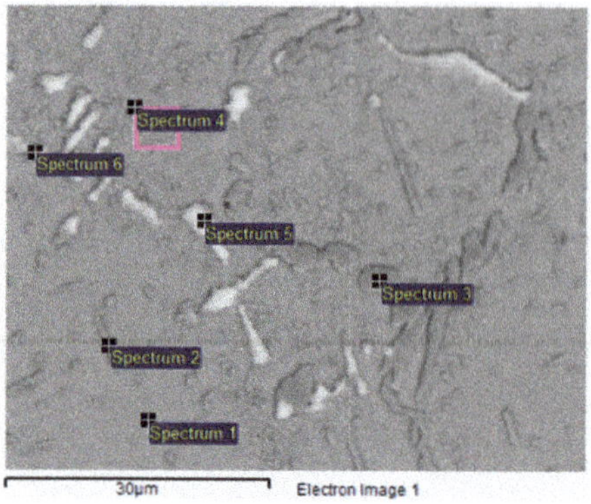

Figure 15. SEM-EDS (BSE) micrograph of the HP40Nb alloy after being overheated at 950 °C/8 h.

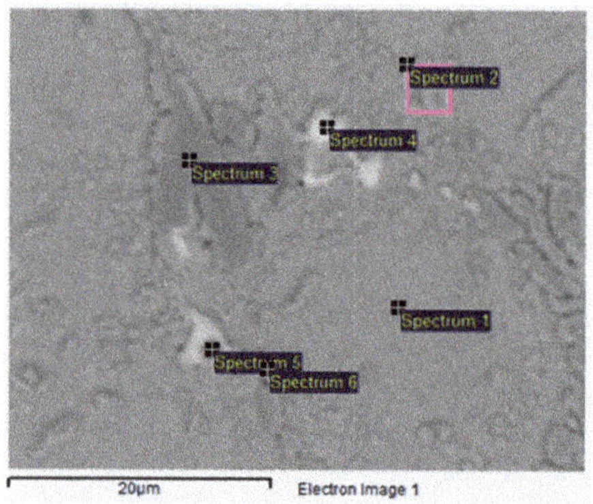

Figure 16. SEM-EDS (BSE) micrograph of the HP40Nb alloy after being overheated at 1050 °C/8 h.

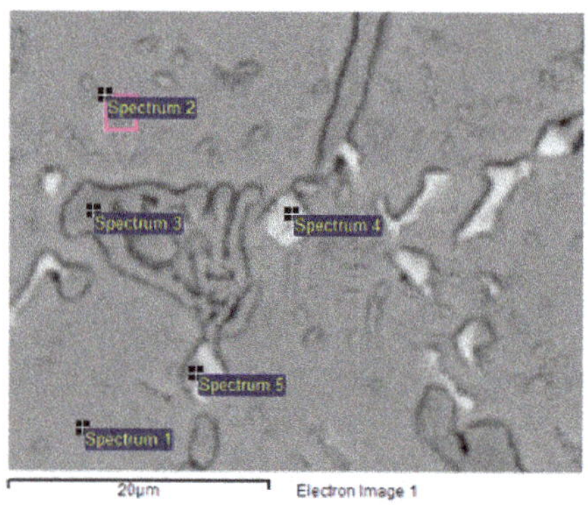

Figure 17. SEM-EDS (BSE) micrograph of the HP40Nb alloy after being overheated at 1150 °C/8 h.

It should be noted that the amount of $M_{23}C_6$-type secondary carbides (M = Cr) formed in the austenitic matrix increased with the increase in overheating temperature to 1050 °C [42], followed by a decrease in the amount of secondary carbides. Secondary carbides precipitate in the austenitic matrix up to a temperature of 1050 °C [19], after which, they undergo partial dissolution (near the primary carbides) or complete dissolution within the dendrites. This change can be observed in detail in SEM Figures 15–17 and in the corresponding EDS Tables 8–10, respectively.

Table 8. Chemical composition of the phases present in the HP40Nb alloy, corresponding to Figure 15.

Spectrum	C	Si	Ti	Cr	Mn	Fe	Ni	Nb	V
Spectrum 1	20.07	2.60		21.62	0.97	28.31	26.45		
Spectrum 2	50.95	1.28		17.41	0.63	15.81	13.92		
Spectrum 3	35.14			52.44	0.98	7.86	3.31		
Spectrum 4	35.60	2.60		19.28	0.90	20.88	20.74		
Spectrum 5	65.32	1.95	0.29	2.5		1.11	4.96	23.65	0.24
Spectrum 6	37.42	4.41	0.15	19.89	0.79	14.65	18.35	4.13	0.21

All results are in atomic%.

Table 9. Chemical composition of the phases present in the HP40Nb alloy, corresponding to Figure 16.

Spectrum	C	Si	Ti	Cr	Mn	Fe	Ni	Nb	V
Spectrum 1	19.38	3.24		18.77	1.22	29.12	28.28		
Spectrum 2	33.71	2.33		17.45	1.02	23.15	22.35		
Spectrum 3	36.62	0.27		45.77	0.95	9.88	6.31		
Spectrum 4	65.05	0.34	0.25	5.10	0.29	5.15	4.56	19.02	
Spectrum 5	45.79		25.97	5.96	0.4	2.72	2.20	16.17	0.78
Spectrum 6	46.56	0.58	18.64	9.07	0.34	8.06	6.93	9.27	0.56

All results are in atomic%.

Table 10. Chemical composition of the phases present in the HP40Nb alloy, corresponding to Figure 17.

Spectrum	C	Si	Ti	Cr	Mn	Fe	Ni	Nb	V
Spectrum 1	21.47	2.81		18.69	1.11	28.13	27.79		
Spectrum 2	31.43	2.28		20.46	0.94	22.91	21.98		
Spectrum 3	36.64			48.06	0.90	9.76	4.63		
Spectrum 4	70.21		2.31	2.65		1.58	1.26		0.26
Spectrum 5	54.73		16.44	3.67		1.09	0.93	22.51	0.63

All results are in atomic%.

Table 8 provides the EDS analysis results of the phases in the sample treated at 950 °C for 8 h, as observed in the SEM micrograph in Figure 15.

The chemical composition (at%) of the phases present in the HP40Nb alloy after being overheated at 1050 °C for 8 h, as observed in Figure 16, is given in Table 9.

Table 10 provides the EDX analysis results of the sample treated at 1150 °C for 8 h, as observed in the SEM micrograph in Figure 17.

Table 11 presents the alloy hardness test results, including the average hardness for each trial. The hardness measurements of the specimens, following overheating treatment at 950, 1050, and 1150 °C, reveal an increase in hardness compared with the initially received specimen. The hardness of the as-received alloy was 215 HV10. The hardness value reached its maximum of 232 HV10 after 8 h at a temperature of 1050 °C. Extending the overheating time from 2 h to 8 h at 1150 °C resulted in an increase in hardness from 207 to 232 HV10.

Table 11. The hardness values of the alloy HP40Nb in its initial state and after overheating at 950, 1050, and 1150 °C for 2 and 8 h, respectively.

Heat Temperature, °C	Hold Time, Hours	Hardness HV10 1	2	3	4	5	Average	Standard Deviation
As-cast	-	228	199	222	230	197	215	16
950	2	210	205	206	206	206	207	2
	8	224	224	230	226	228	226	3
1050	2	207	203	198	206	206	204	4
	8	213	235	235	240	238	232	11
1150	2	204	208	208	209	208	207	2
	8	233	230	230	233	232	232	2

The hardness results are consistent with those reported by other authors [18].

Additionally, Table 11 indicates that the highest hardness values were recorded in specimens subjected to overheating for 8 h.

4. Discussion

The microstructure of the as-received alloy consists of primary carbide networks dispersed within an austenitic matrix, as presented in Figure 1. The dendritic morphology is indicative of a typical cast microstructure. Notably, no secondary carbide formation was observed, as shown in Figure 2. SEM analysis further revealed that the primary carbide networks can be distinguished into two distinct phases, which exhibit dark and bright contrast. As illustrated in Figure 1, the Cr-rich phases identified in the microstructure primarily correspond to $M_{23}C_6$ and M_7C_3 carbides, while the Nb-rich precipitate is characterized as NbC carbide. According to previous investigations [35,43,44], niobium preferentially reacts with carbon during solidification at elevated temperatures, leading to the formation of NbC, with the Nb-to-C atomic ratio playing a crucial role in driving its precipitation. In the present study, the relatively low Nb-to-C ratio (0.2) was insufficient to fully incorporate all available free carbon into NbC [35]. As a result, the remaining carbon in the matrix was preferentially bound by chromium, promoting the formation of Cr-rich carbides at lower solidification temperatures. Due to the high cooling rate, the complete transformation of M_7C_3 into $M_{23}C_6$ was inhibited.

The overheating temperature is an important factor that affects the morphology of carbides. On one hand, irregularly shaped carbides at the grain boundaries hinder the sliding of grain boundaries, while intragranular carbides prevent dislocation movement on the other hand. The precipitation of secondary carbides leads to a reduction in the carbon atom content in the matrix and decreases the effect of solid solution strengthening. During carbide precipitation, the ability of carbides to block dislocation slips is reduced. The morphology of carbides is an important factor influencing the creep properties of the HP40Nb alloy [42,45,46].

In comparison with the microstructure in its initial state, the microstructure of the HP40Nb alloy after overheating at 950 °C for 2 h (Figure 10b) exhibits a markedly higher concentration of carbide in the austenitic region adjacent to the secondary carbide and within the austenitic matrix in the inter-dendritic region. Primary carbides with lamellar and plate-like morphologies are observed along the dendrites, forming discontinuous chains. Additionally, a significant amount of finely precipitated secondary carbides is visible within the austenitic matrix and near the boundaries of the dendrites. On the basis of the XRPD pattern presented in Figure 4, the Cr-rich carbide is identified as an $M_{23}C_6$-type carbide, whereas the Nb-rich phase is classified as an NbC phase.

At an overheating temperature of 1050 °C, in addition to secondary carbides, a quantity of needle-like $M_{23}C_6$ type carbides (where M = Cr, Ni, Fe) was formed within the austenite matrix. The amount of secondary carbides increased as the overheating temperature rose to 1050 °C.

The results of the microstructure analysis of the HP40Nb alloy after overheating at 1050 °C for 2 h are consistent with the findings of Sourmail [27], Fuyang et al. [41], and the work of Martin C. H. Tse [31]. Fuyang et al. [41] also observed the transformation of M_7C_3-type carbides into $M_{23}C_6$ due to the exposure of the HP40Nb alloy to temperatures ranging from 700 to 990 °C for 1 to 6000 h. This transformation is driven by the thermodynamic instability of M_7C_3 at elevated temperatures, resulting in its dissolution and the subsequent precipitation of the more stable $M_{23}C_6$ phase. Furthermore, Tse's research [31] offers valuable insights into the metallurgical advancements in reformer tube alloys, which further enhance the understanding of carbides' evolution under high-temperature conditions.

Liu et al. [17] analyzed the changes in the microstructure and mechanical properties of the HP40Nb alloy during aging at 900 °C from 0 to 2016 h and observed the transformation of M_7C_3-type carbides into $M_{23}C_6$-type carbides.

Almeida et al. [42] studied the influence of niobium content on the microstructural changes in the HP40Nb alloy after aging at 700, 900, and 1100 °C for 1000 h and also observed the transformation of M_7C_3-type carbides into $M_{23}C_6$-type carbides. Wang et al. [35] investigated the effect of overheating the HP40Nb alloy at temperatures ranging from 900 to 1250 °C for 100 h on its microstructure and observed the transformation of M_7C_3-type carbides into $M_{23}C_6$-type carbides.

Zones without secondary carbides appeared in the inter-dendritic space (Figure 10c,d). Near the primary carbides are many finely dispersed secondary carbides in the austenitic matrix.

The quantity of secondary carbides diminished, but their coarsening is evident [34]. Overheating at a temperature of 1150 °C for 2 h resulted in the formation of globular-shaped secondary carbides that separated the regions between the dendrites.

Fuyang et al. [41] also observed the dissolution of secondary carbides with increasing overheating temperature while studying the microstructural changes in the HP series of alloys containing 0.6% C, 1.34% Nb, and Mo during aging at temperatures between 750 and 950 °C across intervals of 1 to 1800 h.

Liu [4] stated that the carbide shape present in the microstructure of the tested material is highly sensitive to the overheating temperature. The overheating of the alloy accelerates the dissolution of smaller carbide particles and the enlargement of larger ones through coagulation, accompanied by morphological changes in carbides that ultimately result in material failure. Coagulation requires a difference in the particle size of the carbide phase, resulting in the solid austenite solution becoming undersaturated and affecting smaller particles under these conditions.

As a result, the smaller particles dissolve, while carbon atoms and alloying elements are transferred via diffusion and incorporated on the surface of the larger particles.

According to the XRPD pattern of the phases present, it can be concluded that the phases present in the original material are the Cr-rich carbide M_7C_3 and the Nb carbide NbC; after overheating, the phases present in the material are Cr-rich carbide $M_{23}C_6$ and the Nb-rich carbide NbC. The difference between the Cr-rich phase in the original material and the superheated material is that the furnace tube is manufactured using a centrifugal casting process. The cooling rate of the original material is relatively fast during the manufacturing process. The metastable M_7C_3 carbide that crystallizes first does not have time to be converted into $M_{23}C_6$ carbide. After overheating, the metastable M_7C_3 carbide becomes stable. The state M_7C_3 carbide is completely transformed into the stable $M_{23}C_6$ carbide [35]. The phase type, grain size, and degree of dispersion of secondary carbides in the microstructure affect the mechanical properties of the material [20].

The difference between samples in the content of intragranular carbide and the degree of grain boundary carbide coarsening increases with longer superheating times.

After exposure for 8 h at 1050 °C (high temperature), secondary chromium carbides precipitate inside the grains (Figure 14b). The appearance of needles (carbides), indicating carbides (dark gray) with gray edges, white carbides, and small gray carbides at the base, can be noticed in Figure 14c. Fewer secondary carbides are present, and these have solidified. The change in the carbide contents can be attributed to the dynamic equilibrium between the dissolution and growth of carbides and the carbon diffusion.

As the temperature rises to 1150 °C for 8 h, the content of secondary carbides decreases due to carbide dissolution. The appearance of carbide needles becomes noticeable; dark gray carbides with gray edges are evident. Small, tiny white carbides are found at the base (as seen in Figure 14c).

Liu and Chen [17] also reported that the content of inter-dendritic carbides increases with longer aging times at a constant temperature of 900 °C.

Extending the overheating time to 8 h at 1150 °C (Figure 14c) results in a reduction in the quantity of secondary carbides and the formation of carbide-free zones within the microstructure. The results indicate that with a prolonged soaking time, chromium carbide decomposes, and the carbon combines with niobium, leading to an increase in niobium-rich carbides. These findings are consistent with the study by Wang et al. [4], which found that the carbide content decreased at a temperature of 1250 °C.

According to the microstructure results of the HP40Nb alloy samples after all overheating treatments (Figures 10 and 14) and the hardness measurement results (Table 11), it can be concluded that the presence of a large number of secondary carbides in the austenitic matrix contributes to the increased hardness of the alloy, as observed by other authors [39,41].

Lanz et al. [39] investigated the effect of overheating temperatures ranging from 750 to 950 °C over various time intervals (1–1800 h) on a modified HP alloy containing 0.6% S and 1.34% Nb. They observed an increase in the alloy's hardness after aging due to the precipitation of needle-like secondary carbides. However, as the aging time and temperature increased, the coarsening of these carbides led to a reduction in hardness. Picasso et al. [33] studied a nickel-based alloy subjected to aging treatments at temperatures between 750 and 900 °C for durations of 1000–4000 h. Their research indicated an increase in hardness values due to aging compared with the as-cast state. However, when the aging time exceeded 500 h, the hardness value began to decline.

The maximum hardness observed in the sample treated at 1150 °C for 8 h is attributed to the dissolution of secondary chromium carbides and the diffusion of carbon into the austenitic matrix.

5. Conclusions

The microstructural evolution of the HP40 alloy alloyed with Nb following short-term overheating was investigated. The main conclusions are based on the experimental results observed.

The as-cast microstructure of the HP40Nb alloy consists of an austenitic dendritic matrix strengthened by a network of primary eutectic-like carbides rich in chromium and niobium of the $M_{23}C_6$ and MC types, respectively.

Microstructural analysis of the HP40 alloy micro-alloyed with Nb after short-term overheating at 950 °C for 2 h revealed a significant presence of secondary carbides in the austenite matrix and near the primary carbides.

Exposure to a temperature of 1050 °C for 2 h resulted in a significant quantity of secondary carbides being dispersed and distributed within the austenitic matrix and agglomerated along the grain boundaries. When the overheating time reached 8 h at a temperature of 1050 °C, a finer grain structure was observed, with carbides dissolved along the austenitic grain boundaries.

Increasing the temperature of overheating from 950 °C to 1150 °C resulted in a reduction in secondary carbides within the structure. Additionally, this increase in temperature altered the morphology of the primary carbides, causing them to adopt a spherical shape.

At a higher overheating temperature of 1150 °C, soaking for 2 h yielded a significantly reduced quantity of the secondary carbides within the austenitic matrix and in the surroundings of the primary carbides. Extending the overheating time to 8 h further decreased the number of secondary carbides.

The increase in hardness values is due to the high content of secondary carbides in the austenitic matrix. This increase is also influenced by the dissolution of secondary carbides and the diffusion of carbon within the austenitic structure.

The microstructure studied via the given combination of techniques has provided new insight into how different carbide types and morphologies form during overheating and further highlighted the significance of microstructure control and monitoring in utilization of reformer tubes in extreme conditions.

In general, this research is beneficial, as it provides insights into how short-term high-temperature exposure affects HP40Nb alloys, particularly under real operating conditions.

Author Contributions: Conceptualization, M.T., D.R. and O.E.C.; methodology, O.E.C., P.J., D.R. and A.K.; validation, M.T.; formal analysis, M.R.; investigation, M.T., P.J., A.K. and D.R.; resources, P.J., S.S. and M.R.; data curation, A.K.; writing—original draft preparation, O.E.C. and S.S.; writing—review and editing, D.R., A.K., O.E.C. and P.J.; visualization, M.T. and P.J.; supervision, D.R.; project administration, P.J. and S.S.; funding acquisition, M.R. and S.S. All authors have read and agreed to the published version of the manuscript.

Funding: This work was supported by the Ministry of Science, Technological Development, and Innovation of the Republic of Serbia; Contract No. 451-03-137/2025-03/200108, Faculty of Mechanical and Civil Engineering in Kraljevo, University of Kragujevac; 451-03-136/2025-03/200213, Innovation Centre of the Faculty of Mechanical Engineering, Belgrade, University of Belgrade; 451-03-136/2025-03/200017 through the realization of research theme 1702514 by the "Vinča" Institute of Nuclear Sciences, National Institute of the Republic of Serbia, University of Belgrade; 451-03-137/2025-03/200156 and 01-50/295 by the Faculty of Technical Science, University of Novi Sad; and 451-03-136/2025-03/200126 by the Faculty of Mining and Geology, University of Belgrade, Serbia. The work titled "Insight into the Microstructure Analysis of HP Austenitic Heat-Resistant Steel under Short-Term High-Temperature Exposure" primarily aligns with Sustainable Development Goal (SDG) 9 Industry, Innovation and Infrastructure from the UN's Agenda 2030.

Data Availability Statement: The original contributions presented in the study are included in the article. Further inquiries can be directed to the corresponding authors.

Conflicts of Interest: The authors declare no conflicts of interest.

References

1. *ASTM A297/ASTM A297-21a*; Standard Specification for Castings, Austenitic, for High Temperature Service. ASTM International: West Conshohocken, PA, USA, 2021.
2. Bonaccorsi, L.; Guglielmino, E.; Pino, R.; Servetto, C.; Sili, A. Damage analysis in Fe–Cr–Ni centrifugally cast alloy tubes for reforming furnaces. *Eng. Fail. Anal.* **2014**, *36*, 65–74. [CrossRef]
3. Ilman, M.; Kusmono, N. Analysis of material degradation and life assessment of 25Cr–38Ni–Mo–Ti wrought alloy steel (HPM) for cracking tubes in an ethylene plant. *Eng. Fail. Anal.* **2014**, *42*, 100–108. [CrossRef]
4. Wang, W.Z.; Xuan, F.Z.; Wang, Z.D.; Liu, C.J. Effect of overheating temperature on the microstructure and creep behavior of HP40Nb alloy. *Mater. Des.* **2011**, *32*, 4010–4016. [CrossRef]

5. Tillack, D.J.; Guthrie, J.E. *Wrought and Cast Heat-Resistant Stainless Steels and Nickel Alloys for the Refining and Petrochemical Industries*; Technical Series; Nickel Development Institute: Toronto, ON, Canada, 1998; Volume 10, pp. 71–85.
6. Allahkaram, S.R.; Borjali, S.; Khosravi, H. Investigation of weldability and property changes of high-pressure heat-resistant cast stainless steel tubes used in pyrolysis furnaces after a five-year service. *Mater. Des.* **2012**, *33*, 476–484. [CrossRef]
7. Shen, L.M.; Gong, J.M.; Jiang, Y.; Geng, L.Y. Effects of aging treatment on microstructure and mechanical properties of Cr25Ni35Nb and Cr35Ni45Nb furnace tube steel. *Acta Metall. Sin. Engl. Lett.* **2011**, *24*, 235–242. [CrossRef]
8. Nishiyama, Y.; Otsuka, N.; Nishizawa, T. Carburization Resistance of Austenitic Alloys in CH_4-CO_2-H_2 Gas Mixtures at Elevated Temperatures. *Corrosion* **2003**, *59*, 688–700. [CrossRef]
9. Mobaraki, M.; Afshang, B.; Rahimpour, M.R.; Bahrololoom, M.E. Effect of cracking feedstock on carburization mechanism of cracking furnace tubes. *Eng. Fail. Anal.* **2020**, *107*, 104216. [CrossRef]
10. Tawancy, H.M. Degradation of mechanical strength of pyrolysis furnace tubes by high-temperature carburization in a petrochemical plant. *Eng. Fail. Anal.* **2019**, *16*, 2171–2178. [CrossRef]
11. *ASTM A608/A608M-20*; Standard Specification for Centrifugally Cast Iron-Chromium-Nickel High-Alloy Tubing for High-Temperature Service. ASTM International: West Conshohocken, PA, USA, 2020.
12. Lissarrague, M.H.S.; Lanz, C.A. NbC transformation during aging in HP40-Nb heat resistant alloy. *Acta Metall. Slovaca* **2022**, *28*, 147–150. [CrossRef]
13. Timotijević, M.; Erić Cekić, O.; Rajnović, D.; Dojčinović, M.; Janjatović, P. Microstructural evolution and mechanical properties degradation HPNb alloy degradation after eleven year serviced. *Struct. Integr. Life* **2022**, *22*, 299–304.
14. Zhou, Y.; Liu, Y.; Zhou, X.; Liu, C.; Yu, J. Precipitation and hot deformation behavior of austenitic heat-resistant steels: A review. *J. Mater. Sci. Technol.* **2017**, *33*, 1448–1456. [CrossRef]
15. Dessolier, T.; McAuliffe, T.; Hamer, W.J.; Hermse, C.G.M.; Britton, T.B. Effect of high temperature service on the complex through-wall microstructure of centrifugally cast HP40 reformer tube. *Mater. Charact.* **2021**, *177*, 111070. [CrossRef]
16. Ghatak, A.; Robi, P.S. Investigation of Microstructure and Creep Life Analysis of Centrifugally Cast Fe-Cr-Ni Alloy Reformer Tubes. *Manuf. Sci. Technol.* **2015**, *3*, 155–159. [CrossRef]
17. Liu, C.J.; Chen, Y. Variations of the microstructure and mechanical properties of HP40Nb hydrogen reformer tube with time at elevated temperature. *Mater. Des.* **2011**, *32*, 2507–2512. [CrossRef]
18. Timotijević, M.; Erić Cekić, O.; Rajnović, D.; Janatović, P. Microstructural analysis of a HP 40Nb alloy aged. *Eng. Today* **2022**, *1*, 41–47. [CrossRef]
19. Hetzner, D.W.; Geertruyden, W.V. Crystallography and metallography of carbides in high alloy steels. *Mater. Charact.* **2008**, *59*, 825–841. [CrossRef]
20. Skindaras, R.; Valiulis, A.V.; Spychalski, W.L. The structure and mechanical properties of the high chromium and nickel content cast alloy after long duration work in high temperature. *MECH* **2014**, *19*, 706–710. [CrossRef]
21. Kondo, Y.; Sakurai, Y.; Namekata, J.; Tanaka, M.; Hangai, F. Effect of High Temperature Aging on Eutectic Carbide Morphology of Centrifugally Cast HK40 and HP Steels. *Tetsu Hagane* **1990**, *7*, 1195–1201. [CrossRef]
22. Haidemenopoulos, G.N.; Zervaki, A.D.; Kamoutsi, H.; Polychronopoulou, K. Creep rupture in HP-Nb refractory steel tubes due to short-term overheating. *Eur. J. Mater.* **2021**, *1*, 1–22. [CrossRef]
23. Voicu, R.; Eric, A.; Poquillon, D.; Furtado, J.; Lacaze, J. Microstructure evolution of HP40-Nb alloys during aging under air at 1000 °C. *Mater. Charact.* **2009**, *60*, 1020–1027. [CrossRef]
24. Buchanan, K.G.; Kral, M.V.; Bishop, C.M. Crystallography and Morphology of MC Carbides in Niobium-Titanium Modified As-Cast HP Alloys. *Metall. Mater. Trans. A* **2014**, *45*, 3373–3385. [CrossRef]
25. Bettanini, A.M.; Ding, L.; Mithieux, J.D.; Parrens, C.; Idrissi, H.; Schryvers, D.; Delannay, L.; Pardoen, T.; Jacques, P.J. Influence of $M_{23}C_6$ dissolution on the kinetics of ferrite to austenite transformation in Fe-11Cr-0.06C stainless steel. *Mater. Des.* **2019**, *162*, 362–374. [CrossRef]
26. Zhang, J.; Li, J.; Shi, C.; Li, J. Evolution of Eutectic Carbide during M_7C_3 /$M_{23}C_6$ in situ Transformation in Martensitic Stainless Steel. *Steel Res. Int.* **2022**, *93*, 2200231. [CrossRef]
27. Sourmail, T. Precipitation in creep resistant austenitic stainless steel. *Mater. Sci. Technol.* **2001**, *17*, 1–13. [CrossRef]
28. Nascimento, M.L.C.; da Cruz Gallo, F.; dos Santos Queiroz, F.; Mendes, M.C.; Eckstein, C.B.; Nogueira, L., Jr.; Le May, I.; Ribeiro Pereira, G.; de Almeida, L.H. Effect of short-time overheating in the morphology of primary carbides network in Nb and NbTi-modified HP stainless steels steam reforming tubes. *J. Mater. Res. Technol.* **2023**, *22*, 382–392. [CrossRef]
29. Ma, Y.W.; Yang, G.; Yoon, K.B.; Le, T.G. Microstructure Evaluation During Short Term Creep of Cr35Ni45Nb Cast Alloy Reformer Tube. *Met. Mater. Int.* **2020**, *27*, 5165–5172. [CrossRef]
30. Haidemenopoulos, G.N.; Polychronopoulou, K.; Zervaki, A.D.; Kamoutsi, H.; Alkhoori, S.I.; Jaffar, S.; Cho, P.; Mavros, H. Aging Phenomena during In-Service Creep Exposure of Heat-Resistant Steels. *Metals* **2019**, *9*, 800. [CrossRef]
31. Tse, M.C.H.; Livera, E.R.; Christofidou, K.A. Metallurgical developments in steam-methane reformer tube alloys. *Mater. Sci. Technol.* **2024**, 1–15. [CrossRef]

32. *ISO 6507-1:2018*; Metallic Materials—Vickers Hardness Test—Part 1: Test Method. International Organization for Standardization: Geneva, Switzerland, 2018.
33. Picasso, A.C.; Lanz, C.A.; Lissarrague, M.S.; Garófo, A.D. Microstructure Evolution of a Nickel-Base Alloy Resistant to High Temperature during Aging. *JMMCE* **2016**, *4*, 48–61. [CrossRef]
34. Maminska, K. Optimisation microstructuraled'unacier HP pour des applications àhaute temperature. Ph.D. Thesis, Ecole Nationale Supérieuredes Minesde Saint-Etienne, Saint-Etienne, France, 2013. Available online: https://theses.hal.science/tel-02003637/ (accessed on 1 January 2020).
35. Zhou, Y.; Fuyang, C.; Gong, J.; Geng, L.; Jiang, Y.; Tang, J. Effect of overheating on the microstructures and mechanical properties of HP 40-Nb furnace tube. *J.Press. Vessel Technol.* **2021**, *38*, 1–9. [CrossRef]
36. Wang, M.; Flahaut, D.; Zhang, Z.; Jones, I.P.; Chiu, Y. Primary carbide transformation in a high performance micro-alloy at 1000 °C. *J. Alloys Compd.* **2019**, *781*, 751–760. [CrossRef]
37. Ray, A.K.; Kumar, S.; Krishna, G. Microstructural studies and remnant life assessment of eleven years service exposed reformer tube. *Mater. Sci. Eng. A* **2011**, *529*, 102–112. [CrossRef]
38. Liu, W.; Zheng, Y.G.; Yao, Z.M.; WU, X.Q.; Ke, W. Cavitation Erosion of 20SiMn and 0Cr13Ni5Mo Steels in Distilled Water with and without Sand. *Acta Metall. Sin.-Engl. Lett.* **2001**, *37*, 197–201. [CrossRef]
39. Lanz, C.; Brizuela, G.; Juan, A.; Simonetti, S. Microstructural evolution of a modified HP alloy: Experimental and complementary computational study. *J. Adv. Phys.* **2017**, *13*, 5141–5145. [CrossRef]
40. Delić, A.; Oruč, M.; Rimac, M.; Gigović-Gekić, A.; Sunulahpašić, R. The influence of solution annealing on microstructure and mechanical properties of heat resistant cast steel HK30 modified by niobium. *Metall. Mater. Eng.* **2009**, *25*, 237. [CrossRef]
41. Fuyang, C.M.; Chen, J.; Shao, B.; Zhou, Y.; Gong, J.; Guo, X.; Jiang, Y. Effect of microstructural evolution in thermal exposure on mechanical properties of HP40Nb alloy. *Int. J. Press. Vessel. Pip.* **2021**, *192*, 104391. [CrossRef]
42. Almeida, L.H.D.; Ribeiro, A.F.; May, I.L. Microstructural characterization of modified 25Cr–35Ni centrifugally cast steel furnace tubes. *Mater. Charact.* **2003**, *49*, 219–229. [CrossRef]
43. Shi, S.; Lippold, J.C. Microstructure evolution during service exposure of two cast, heat-resisting stainless steels—HP-Nb modified and 20-32Nb. *Mater. Charact.* **2008**, *59*, 1029–1040. [CrossRef]
44. Lissarrague, M.H.S.; Limandri, S.; Prado, F.A.; Picasso, C. Study of the Microstructural Evolution in a 35Ni-25Cr-Nb Heat-Resistant Alloy by Dilatometry and Electron Microscopy. *Metallogr. Microstruct. Anal.* **2018**, *7*, 356–362. [CrossRef]
45. Ghatak, A.; Robi, P.S. Effect of Microstructure with Hardness on Heat Treatment of HP40Nb Microalloyed Reformer Tube. In Proceedings of the 5th International & 26th All India Manufacturing Technology, Design and Research Conference (AIMTDR 2014), Guwahati, India, 12–14 December 2014.
46. Gao, Y.J.; Yang, Y.; Yao, F.; Ye, C.; Yi, Z.D.; Ma, S. Effect of tungsten on the microstructure evolution and mechanical properties of yttrium modified HP40Nb alloy. *Mater. Sci. Eng. A* **2011**, *529*, 361–369. [CrossRef]

metals

MDPI

Article

Indentation-Free Resistance Spot Welding of SUS301L Stainless Steel

Yutong Liu [1], Yuming Xie [1,2,*], Xiuwen Sun [1], Licheng Sun [1], Naijie Wang [2], Xiaotian Ma [2], Xiangchen Meng [1,2] and Yongxian Huang [1,2,*]

[1] State Key Laboratory of Precision Welding and Joining of Materials and Structures, Harbin Institute of Technology, Harbin 150001, China; 2021110889@stu.hit.edu.cn (Y.L.); 23b909132@stu.hit.edu.cn (X.S.); sunlicheng@sjtu.edu.cn (L.S.); mengxch@hit.edu.cn (X.M.)

[2] Zhengzhou Research Institute, Harbin Institute of Technology, Zhengzhou 450046, China; wangnaijie@hit.edu.cn (N.W.); maxiaotian@hit.edu.cn (X.M.)

* Correspondence: ymxie@hit.edu.cn (Y.X.); yxhuang@hit.edu.cn (Y.H.)

Abstract: Paint-free bodywork has become an attractive alternative for rail vehicles, in the direction of easy maintainability and low manufacturing costs. However, conventional resistance spot welding inevitably leaves indentation marks to detrimentally reduce the optical homogeneity of the paint-free bodywork. In light of this, indentation-free resistance spot welding is proposed for joining SUS301L stainless steel sheets in order to achieve superior surficial integrity. A tiny SUS301L steel ball with a diameter of 1.5 mm was chosen as the intermediate filler between two steel sheets to avoid the formation of surficial indentation. The influence of welding current and welding time on the mechanical properties of joints was studied. The optimal parameters of the mechanical properties were obtained when the welding current was 8.0 kA, the welding time was 150 ms, the electrode pressure was 0.35 MPa, and the electrodes were cylindrical planar electrodes, which was determined by comparing the tensile shear test results. The surficial indentation depth was less than 1% of the plate thickness, and no observable indentations were seen on the surface of the optimized welding spots.

Keywords: stainless steel; resistance spot welding; indentation-free welding; mechanical properties

1. Introduction

Paint-free bodywork offers advantages, such as environmental protection, easy maintainability, and low manufacturing costs, making it an appealing alternative for rail vehicles [1]. SUS301L is currently the most commonly used material for rail vehicles and automobile bodies due to its attractive appearance, paint-free properties, anti-corrosion capabilities, and lightweight nature [2,3]. One of the significant technical challenges in achieving paint-free bodywork is indentation at the weld joints [4].

Resistance spot welding is extensively utilized in the welding of steel sheet structures for automobiles and rail vehicles because of its brief welding duration, uncomplicated process, and high level of automation [5–7]. More than 90% of assembly work in a car body is completed by resistance spot welding at present [8–10], and there are 2000 to 5000 spots made by resistance spot welding on each car in rail vehicles [11,12]. While ensuring the safety and reliability of joints, it is particularly important to study the surface quality of the spot-welded joints of paint-free stainless steel rail vehicles [13–15]. An obvious indentation will always be left on the surface of the thin sheet due to the technical characteristics of resistance spot welding, which not only affects the aesthetics of the body but also reduces the corrosion resistance of the joints [16,17]. The realization of indentation-free resistance spot welding of SUS301L stainless steel is of great significance for the production of paint-free rail vehicles.

The welding current, welding time, and electrode pressure are three key process parameters that affect the joint quality of the resistance spot welding process [18–20]. Sreehari et al. [21] found, concerning the indentation levels, that welding time has a lesser influence when compared to welding current and electrode force. Gong et al. [22] found that increasing compressive residual stress can improve the brittle fracture resistance and corrosion resistance of metal materials. Han et al. [23] explored the effect of the indentation on the sizing of the spot weld nugget. The results show that the existence of an indentation may result in under-sizing of the nugget. Pouranvari et al. [24] studied the mechanical performance of dissimilar resistance spot welding between DP600 and AISI 1008 low-carbon steels, concluding that increases in welding current and welding time result in increases in electrode indentation depth. Rajesh et al. [25] used the Taguchi approach to optimize the control parameters of the dual pulse resistance welding of 2 mm ASTM A240 stainless steel sheets. The findings indicate that the magnitude of the second pulse current is the most important component that influences indentation in dual pulse resistance spot welding. Qi et al. [26] designed a magnetically assisted resistance spot welding (MA–RSW) device for unilateral surface indentation of the traditional stainless steel RSW joint. The welding penetration rate was reduced by using single-sided MA–RSW at the permanent rate, thus the thermal influence on the base metal surface was reduced, and the surface indentation depth was reduced by more than 45% compared with the traditional process. Xia et al. [27] designed a robotic servo gun and an accompanying displacement monitoring system. Joint quality can be measured online and the welding parameters can be adjusted through this system to reduce the impact of surface indentation. These studies show the factors and control methods that affect the indentation of resistance spot welding, but they cannot achieve complete indentation-free spot welding.

In this paper, we propose a novel indentation-free resistance spot welding process. A tiny steel ball was chosen as the intermediate filler to avoid the formation of surficial indentation. The surface indentation, nugget morphology, microstructure, and macroscopic mechanical properties of the joint were investigated in detail.

2. Materials and Methods

Respectively, 0.6-mm-thick and 1.5-mm-thick SUS301L stainless steel sheets (Hebei Jingche Railway Vehicle Equipment Co., Ltd., Baoding, China) were used as upper sheets and lower sheets. The thin sheets were more easily deformed and released stress, so they were used to reduce the deformation of thick sheets. The intermediate filler was a SUS301L stainless steel ball with a diameter of 1.5 mm. The composition and mechanical properties of SUS301L are listed in Tables 1 and 2. Both the upper and lower electrodes used in spot welding are cylindrical planar electrodes with a diameter of 20 mm. Electrodes were connected to a water-cooling device for rapid cooling.

Table 1. Composition of SUS301L stainless steel (wt.%).

Element	C	Si	N	Mn	P	S	Ni	Cr	Fe
Mass fraction	0.02	0.50	0.10	1.0	0.03	0.006	7.00	17.00	Bal.

Table 2. Mechanical properties of SUS301L stainless steel.

Material	Ultimate Tensile Strength (MPa)	Yield Strength (MPa)	Elongation (%)
SUS301L	930	685	20

A resistance spot welding process for thin and thick plates was developed to reduce surface indentation. The process flow is illustrated in Figure 1. A tiny steel ball was initially welded onto the thin plate using a small welding current, welding time, and electrode pressure, followed by welding between the two sheets. Choosing the right size ball was important for the welding process. The large diameter of the ball caused a large gap between the two plates, resulting in low joint strength. The steel ball with a very small diameter melted instantly and was not able to work. In this study, we chose steel balls with a diameter of 1.5 mm.

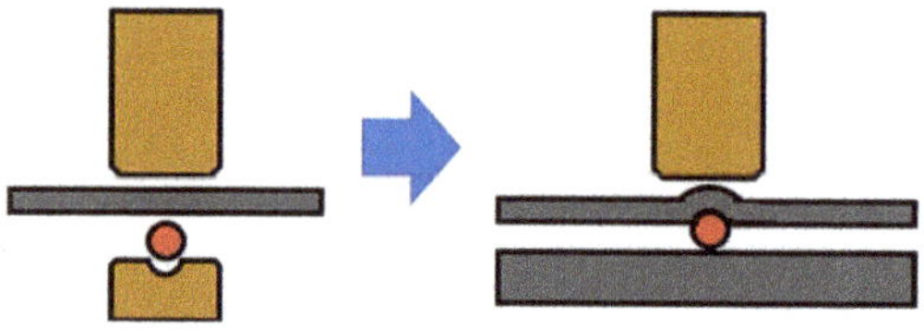

Figure 1. Process of indentation-free resistance spot welding.

Figure 2 shows the principle of the indentation-free resistance spot welding process. This method achieved low current density (J) at the surface of the base metal in contact with the electrode, while concentrating current at the middle layer of the steel ball. According to Ampere's law and resistance law, the resistance heat generated by current is related to resistivity (ρ), cross-sectional area (S), current density, and time (t). The calculation formula for resistance heat generated by current flowing through a conductor per unit thickness is as follows:

$$Q = \rho J^2 S t$$

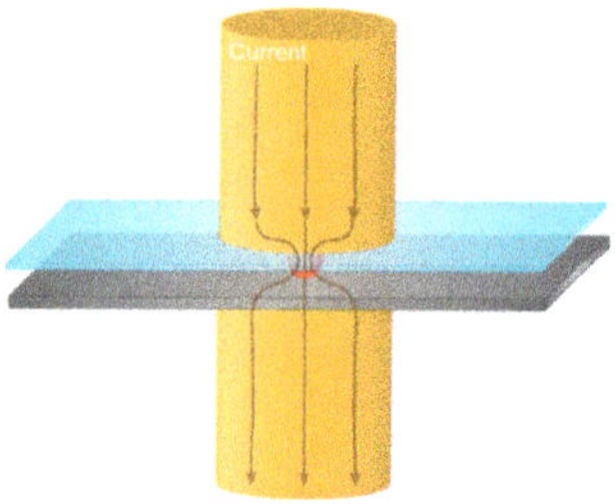

Figure 2. Principle of indentation-free resistance spot welding.

Large-diameter electrodes were used to increase the contact area between the electrode and the steel plate. The position of the steel ball has a much smaller cross-sectional area than other sites. A high current density in the middle layer of the steel ball generated significant resistance heat, causing the steel ball to melt and form a spot welding nugget. The resistance heat generated by low current density on the surface of the welded base metal in contact with the electrode was minimal. The surface indentation, ablation, and welding deformation were reduced in this way.

A copper electrode with a pit was utilized to position the ball during spot welding of the ball and sheet. The welding current was set at 1.5 kA, with a welding time of 150 ms, and an electrode pressure of 0.10 MPa.

Welding spatter was prone to occur due to the high current density at the steel ball when welding two sheets. To solve this problem, a preheating input of 2 kA welding current, 100 ms welding time, and electrode pressure of 0.35 MPa was applied before welding two sheets. According to preliminary trials, the electrode pressure was set at 0.35 MPa. Welding current and welding time were selected for optimization. We set the welding current at 7.0 kA, 7.5 kA, 8.0 kA, and 8.5 kA and the welding time at 120 ms, 150 ms, 180 ms, and 210 ms.

The metallographic and tensile shear test samples were prepared perpendicularly to the cross-sections of the resistance spot-welding joints by electric wire cutting. The metallographic samples were ground, polished, and etched with a mixture of hydrofluoric acid and nitric acid for 7 min. The cross-sectional morphology, microstructure, and surface indentation of each zone of the sample were observed using a VHX-1000E optical microscope (Keyence, Osaka, Japan). A KC-X1000 laser confocal microscope (Kathmatic, Nanjing, China) was used to obtain the contour of the joint surface, and a straight line on the indentation was selected to measure the indentation contour.

The tensile shear sample size is 50 mm × 15 mm, as shown in Figure 3. The tensile shear test was carried out on an electronic universal material test machine at ambient temperature with a tensile speed of 2 mm/min. The specimens were grasped with shims that were equal in thickness to the sheets. The typical fracture was observed and analyzed using a Zeiss Merlin Compact scanning electron microscope.

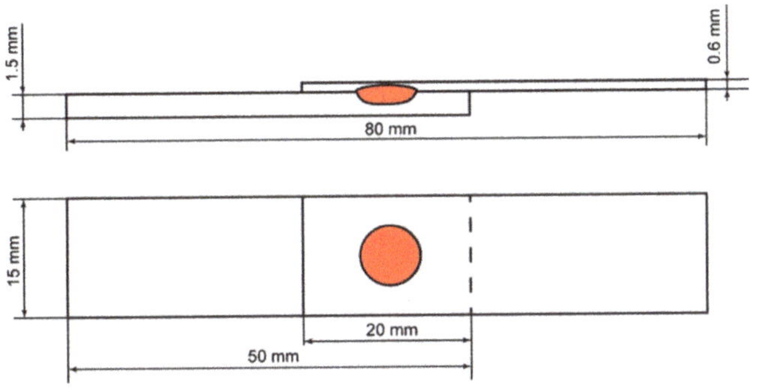

Figure 3. Dimensions of tensile shear test specimens.

3. Results and Discussion

The macroscopic morphology of welding between a 1.5 mm diameter ball and a 0.6 mm thick steel sheet is shown in Figure 4. The steel ball was deformed under the action of welding heat input and electrode pressure, but the outline remained circular. The pitted part of the steel ball did not produce large resistance heat, so it retained the original shape, and played a role in welding between the two sheets.

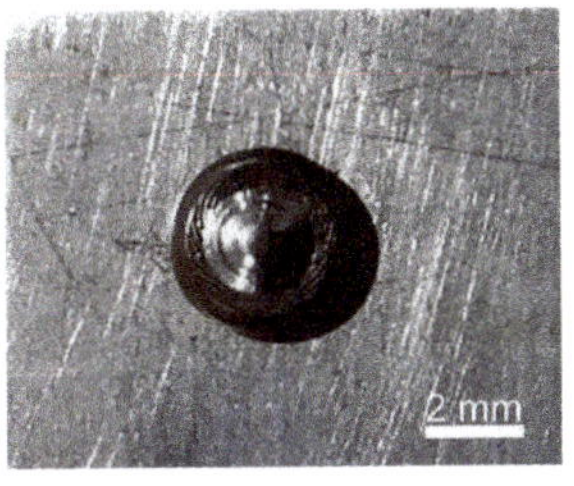

Figure 4. Appearance of 1.5 mm diameter steel ball welded onto 0.6 mm thick sheet.

The macroscopic morphology of several groups of indentation-free resistance spot welding process joints is shown in Figure 5. The surface indentation of the joint was minimal, and there were no ablation marks or color changes on the surface, resulting in a regular circular shape. There was almost no deformation on the side surface of the thick plate. Variations in welding parameters had little impact on the surface indentation and tablation marks of the joint within a certain range. This phenomenon was attributed to the larger contact area between the planar electrode and the thick plate side compared to that between the steel ball and steel plate, leading to relatively low current density and heat input on the thick surface of the plate. As a result, the surface indentation, welding deformation, and ablation marks were small. Changes in welding parameters did not greatly affect the surface morphology of spot-welded joints.

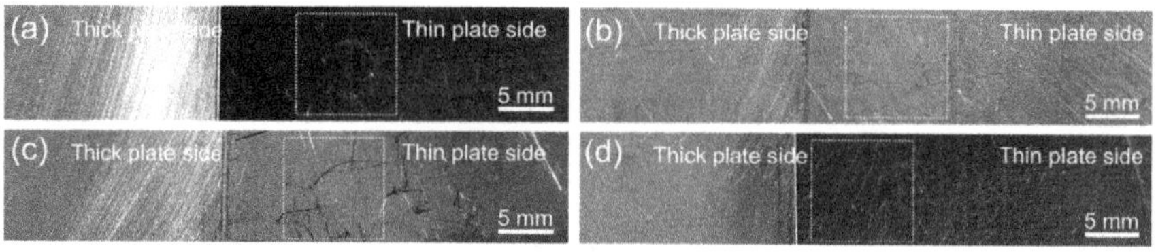

Figure 5. Surface morphology of spot-welded joints: (**a**) 7.0 kA, 150 ms, 0.35 MPa, (**b**) 7.5 kA, 150 ms, 0.35 MPa, (**c**) 8.0 kA, 150 ms, 0.35 MPa and (**d**) 8.0 kA, 180 ms, 0.35 MPa.

Figure 6 shows the nugget geometry of the indentation-free resistance spot-welding joint under different welding parameters, as observed through an optical microscope. The nugget shape of the joint showed a basin shape with a wide upper and narrow lower, which is because the resistance heat generated by the intermediate filler under the action of current constraint was larger than that on the thick plate side. However, because the intermediate filler was also affected by the current density, the welding parameters will greatly impact the spot-welding nugget. In this study, the penetration rate is expressed as the percentage of the thickness of the nugget on the side of the thick plate in proportion to the thickness of the thick plate. The nugget diameter increased from 4.80 mm at 7.5 kA to 5.18 mm at 8.0 kA as the welding current increased, and the penetration rate also increased from 47% to 58%. Welding spatter occurred when the welding current was increased to 8.5 kA. The nugget diameter was 4.07 mm and the penetration rate was 57% at this time, which is smaller than that at 8.0 kA, and the nugget size at the joint interface and the sheet side was also small. This is because the welding spatter caused the heat at the joint interface to be lost as the liquid metal flew away, further leading to insufficient heat input and poor fusion at the joint interface. It is necessary to avoid poor fusion and spatter to ensure the load-bearing capacity of the spot-welded joint. The effect of welding time on nugget morphology is similar to that of welding current. The nugget diameter and penetration rate increased from 4.53 mm and 56% at 120 ms to 5.18 mm and 58% at 150 ms with the increase in welding time.

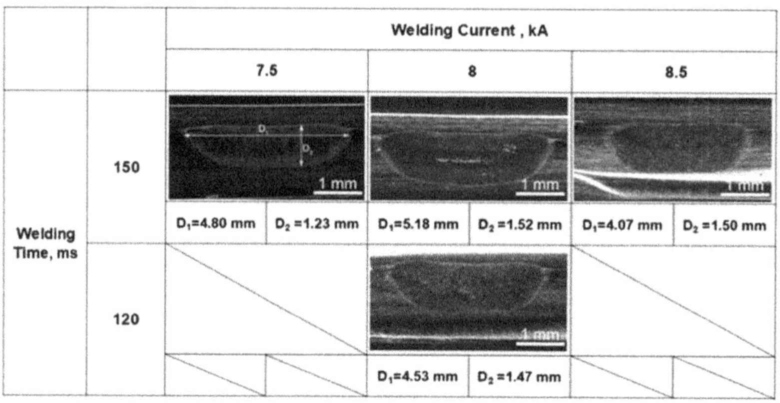

Figure 6. Nugget morphologies of resistance spot weld joint.

Figure 7 shows the microstructure of the resistance spot weld joints (8.0 kA, 120 ms, 0.35 MPa). Typical zones of the joint were the nugget zone and the heat-affected zone. The width of the heat-affected zone was small and was mainly composed of the coarse austenite grains formed by recrystallization, due to the rapid cooling effect of the water in the electrode [28]. This area was considered the weak point of the joint. The nugget zone consisted of columnar austenite grains growing from the thin plate and thick plate to the nugget center, because the temperature gradually decreased from the nugget to both sides.

However, the molten steel ball was extruded towards both sides of the nugget to form an extrusion zone, as shown in Figure 8c. This zone was also classified as a heat-affected zone. Two different forms of the joint were found at the joint interface between the nugget and the thin plate (shown by the yellow line in Figure 7), corresponding to Figure 8a,b, respectively.

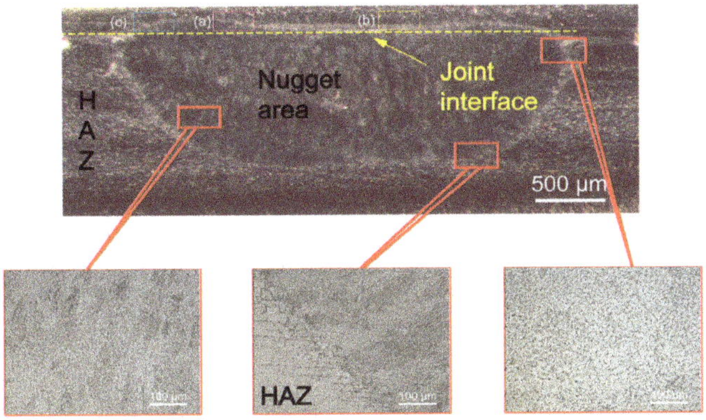

Figure 7. Microstructure of resistance spot-welding joint: (**a**) sampling position of Figure 8a, (**b**) sampling position of Figure 8b, and (**c**) sampling position of Figure 8c.

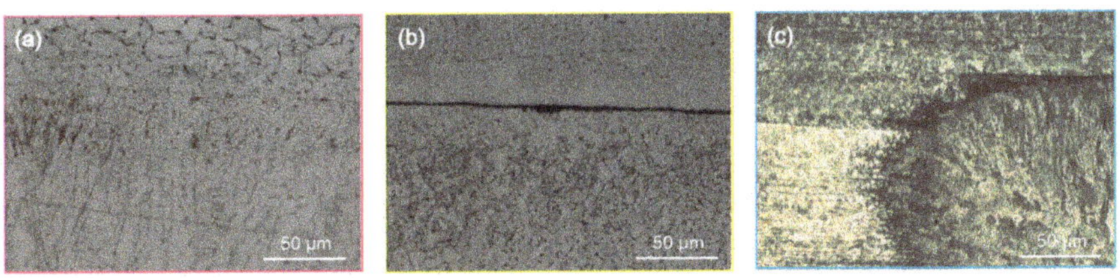

Figure 8. Interfaces of the welded joints: (**a**) no obvious interface, (**b**) distinct interface with cracks, and (**c**) interface with extruded steel ball.

Figure 8a shows the transition form of the normal resistance spot-welding joint from the nugget zone to the heat-affected zone. This transition is characterized by a shift from columnar crystals to coarse equiaxed austenite grains, with no obvious joint interface. As shown in Figure 8b, the area with an obvious joint interface was located in the middle of the nugget, at the position of the pre-welded steel ball of the thin plate. There was an obvious joint interface in this area, which accounted for about one-third of the diameter of the nugget, which seriously affected the mechanical properties of spot-welded joints. The area can be considered as incomplete fusion at the joint interface. According to the location and joint of the zone shown in Figure 8b, the formation of this zone was related to the process of pre-welding steel balls for the sheet. A small welding current and welding time were adopted in the process of welding the sheet and steel balls. Although this can maintain the shape of the steel ball and make it connect with the sheet to a certain extent, the joint was not strong. There was an obvious joint interface between the plate and the steel ball. In the process of base metal welding, the contact resistance between the steel ball and the thick plate generated a large amount of resistance heat at the joint of the steel ball and the thick plate, causing this area to form a good fusion. However, the contact resistance between the steel ball and the thin plate disappeared in the step of welding two sheets. This resulted in insufficient heat input at the position, causing the original joint state to remain unchanged and leading to poor fusion in the interface area. The fracture position of the joint was located at the joint interface between the thin plate and the nugget in the subsequent fracture analysis, which agreed with the above analysis of the cause of the poor fusion.

A set of tests was designed to study the changing trend of the maximum tensile shear of the joint with the welding current. The specific process parameters are shown in Table 3. An increase in welding current led to the highest maximum tensile shear and nugget diameter. However, welding spatter occurred at 8.5 kA, resulting in a decrease in maximum tensile shear.

Table 3. Welding current and welding time test parameters.

Samples	Welding Current	Welding Time	The Highest Value of Tensile Shear Force	Welding Spatter (Yes/No)
1	7.0 kA	150 ms	6027 N	N
2	7.5 kA	150 ms	6052 N	N
3	8.0 kA	150 ms	6414 N	N
4	8.5 kA	150 ms	5530 N	Y
5	8.0 kA	120 ms	6403 N	N
6	8.0 kA	180 ms	6489 N	N
7	8.0 kA	210 ms	5810N	Y

The variation trend in the maximum tensile shear of spot-welded joints was studied under the condition that only the power on time was changed on this basis. The maximum tensile shear force increased with the increase in welding time, but the increase was small. The maximum tensile shear plummeted when the welding time was 210 ms due to the welding spatter.

Figure 9a,b shows the maximum tensile shear force of specimens at various welding currents and the load–displacement curve of the four groups of corresponding parameters. Figure 9c,d shows the variation trend in the maximum tensile shear force with welding time and the load–displacement curves of the four groups of corresponding parameters. The optimal parameters of the mechanical properties were obtained when the welding current was 8.0 kA, the welding time was 150 ms and the electrode pressure was 0.35 MPa. To predict the tensile shear force of resistance spot-welded thin steel sheets, Majlinger et al. [29] recommended a formula based on the materials' tensile strength and the sheet thickness:

$$TSF = -10.10 + 0.0088 \cdot R_m + 15.80 \cdot t \ (kN)$$

In this formula, TSF is the predicted shear tensile strength of the optimized joints, R_m is the tensile strength of the steel in MPa and t is the sheet thickness in mm. The maximum tensile shear force obtained under the optimal parameters is predicted to be 7.564 kN according to this formula. The experimental values were close to the predicted values.

We speculated that the optimal parameters obtained may be mainly influenced by the size of the nugget. With the increase in welding current and welding time, the nugget size increases, and the effective bonding area of the joint also increases. Excessive heat input led to welding spatter and large welding residual stress when the welding current or welding time was too large, and the mechanical properties decreased therefore [30,31].

Figure 10 shows the macro fracture morphology of samples 1–4. The macro fracture interface shown in the figure was the fracture morphology of the thick plate side. The fracture mode of the tensile shear specimen of samples 1, 2, and 4 was the interface fracture mode. The fracture occurred at the position between the thin plate and the nugget (incomplete fusion zone). For the tensile shear specimen of sample 3, the fracture mode of the tension-shear specimen was button fracture mode. The fracture position was in the heat-affected zone on the thin side, and the fracture interface did not pass through the nugget but broke along the heat-affected zone on the sheet side. It is generally considered that the button fracture mode of the joint is better than the interface fracture mode [32]. The extrusion zone of sample 3 (as shown in Figure 10c) and the thin plate side were first separated and formed a smooth fracture interface, indicating that its strength was low.

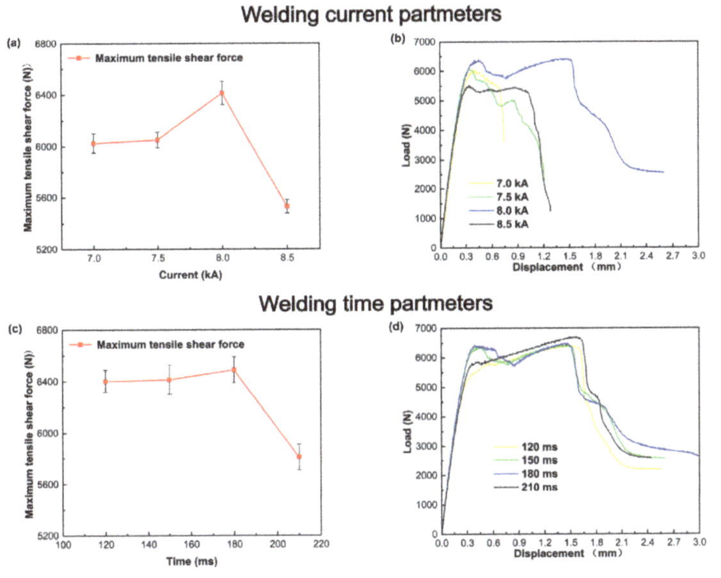

Figure 9. Tensile shear curve with different parameters: (**a**) tensile shear force curve with different welding currents, (**b**) load–displacement curve, (**c**) tensile shear force curve with different welding time, and (**d**) load–displacement curve.

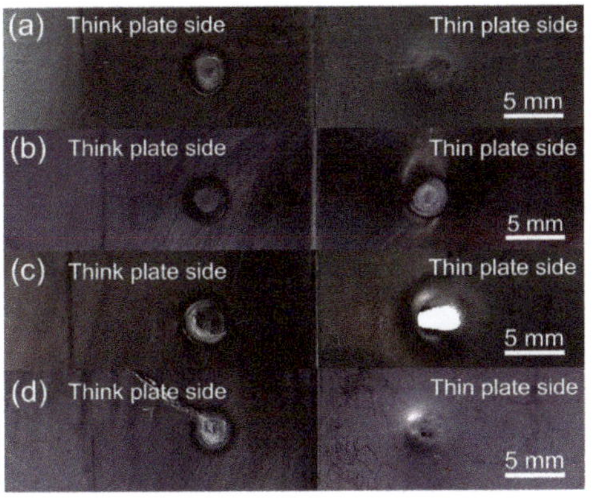

Figure 10. Fracture surface morphology of RSW joint: (**a**) 7.0 kA, 150 ms,0.35 MPa, (**b**) 7.5 kA, 150 ms, 0.35 MPa, (**c**) 8.0 kA, 150 ms, 0.35 MPa, and (**d**) 8.5 kA, 150 ms, 0.35 MPa.

The fracture interface morphology and fracture position of the spot-welded joint are shown in Figure 11. The maximum tensile shear of the button fracture mode samples was larger than that of the interface fracture mode samples. This was consistent with the results of the maximum tensile shear curve shown in Figure 6. The reason was related to the incomplete fusion at the joint of the thin plate and the nugget. The incomplete fusion reduced the bonding strength at the joint interface of the nugget and the thin plate side, so that the interface fracture mode occurred.

Figure 12a,b shows the surface indentation morphology of the RSW joint obtained under the optimal parameters of mechanical properties (8.0 kA, 150 ms and 0.35 MPa). The indentation was shallow and presented a regular circle. Figure 12c shows the indentation contour of the red line segment in Figure 12b. The height difference between the indentation on the joint surface and the base material was small. The maximum indentation height difference was about 25 μm. The maximum indentation depth was about 1% of the plate thickness for the thick plate side with a thickness of 1.5 mm. It can be considered that the surface of the spot welding joint had no indentation under this set of parameters.

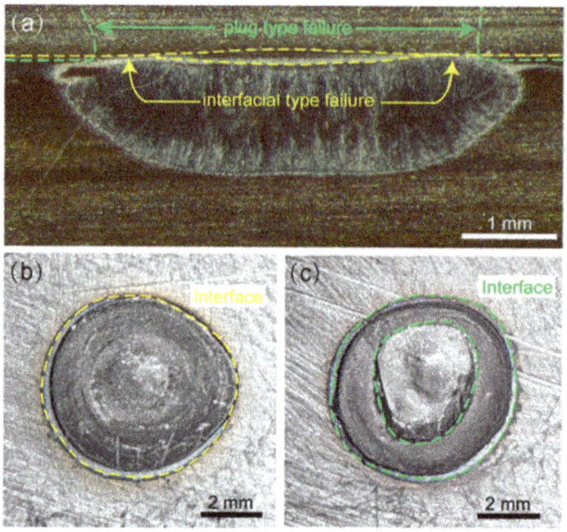

Figure 11. Fracture interface morphologies and fracture locations: (**a**) nugget morphology of the joint, 8.0 kA, 180 ms, 0.35 MPa, (**b**) macroscopic fracture interface of interfacial fracture mode, 7.0 kA, 150 ms, 0.35 MPa, and (**c**) macroscopic fracture interface of button fracture mode, 8.0 kA, 150 ms, 0.35 MPa.

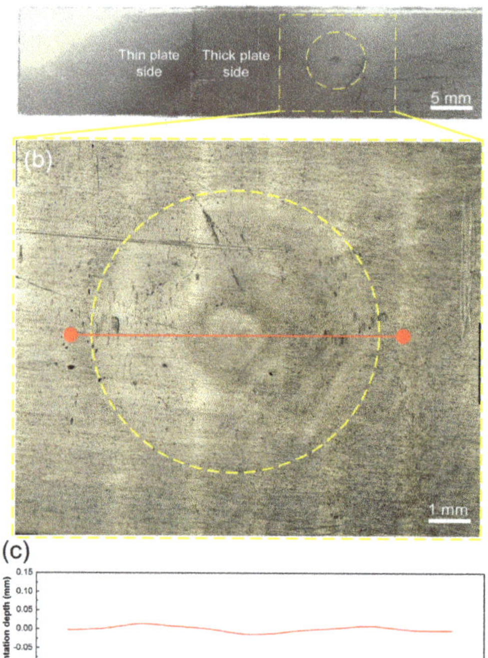

Figure 12. Surface morphologies of the joint. (**a**) Macroscopic indentation on the side surface of the thick plate, (**b**) the thick plate side indentation under an ultra-deep microscope, and (**c**) indentation depth of the red line joint surface.

4. Conclusions

This paper proposed a novel resistance spot welding process through the intermediate layer of steel balls to confine the current and obtain an indentation-free surface on one side of the spot-welding joint. The process changed the current density of different parts, resulting in high current density at the nugget. The resistance heat is concentrated in the intermediate filler. The surface of the plate had a small welding heat input, leading to minimal deformation, indentation, and ablation marks. The highest value of tensile shear force increased with increasing welding current or welding time. Welding spatter occurred when the welding current or welding time was too large, which decreased the highest value of tensile shear force. The microstructure of this joint was similar to that of traditional resistance spot-welding joints. The maximum tensile shear force reached 6414 N when the welding current was 8.0 kA, the welding time was 150 ms, and the electrode pressure was 0.35 MPa, by using cylindrical planar electrodes. The surface indentation depth of the joint under this set of parameters was about 1% of the plate thickness at this time, and almost no indentations could be achieved.

Author Contributions: Conceptualization, Y.L. and Y.X.; methodology, Y.L. and X.S.; validation, L.S.; formal analysis Y.X. and X.M. (Xiaotian Ma); investigation N.W. and X.M. (Xiaotian Ma); writing—original draft preparation Y.L.; writing-reviewing and editing, X.M. (Xiangchen Meng), Y.H. and Y.X. All authors have read and agreed to the published version of the manuscript.

Funding: This work was jointly supported by the National Natural Science Foundation of China (Nos. 52305345, 52175301, 52205350).

Data Availability Statement: The data presented in this study are available on request from the corresponding author. The data are not publicly available due to privacy restrictions.

Conflicts of Interest: The authors declare no conflicts of interest.

References

1. Xu, W.; Zhang, B.; Deng, Y.; Wang, Z.; Jiang, Q.; Yang, L.; Zhang, J. Corrosion of rail tracks and their protection. *Corros. Rev.* **2021**, *39*, 1–13. [CrossRef]
2. Fu, Z.H.; Li, T.; Shan, M.L.; Gou, G.Q.; Zhu, Z.Y.; Ma, C.P.; Gao, W.; Hu, Y.C. Hydrogen atoms on the SCC behavior of SUS301L-MT stainless steel laser-arc hybrid welded joints. *Corros. Sci.* **2019**, *148*, 272–280. [CrossRef]
3. Li, J.; Luo, X.; Ma, G.; Wang, J.; Pan, J.; Ruan, Q. The effect of cold rolled reduction ratios on grain boundary character and mechanical properties of the SUS301L austenitic stainless steel. *Mater. Res. Express* **2019**, *6*, 126587. [CrossRef]
4. Kalyankar, V.D.; Chudasama, G.P. Influence of electrode tip diameter on metallurgical and mechanical aspects of spot welded duplex stainless steel. *High Temp. Mater. Process.* **2020**, *39*, 317–327. [CrossRef]
5. Martín, Ó.; Tiedra, P.D.; López, M.; San-Juan, M.; García, C.; Martín, F.; Blanco, Y. Quality prediction of resistance spot welding joints of 304 austenitic stainless steel. *Mater Des.* **2009**, *30*, 68–77. [CrossRef]
6. Feng, Q.B.; Li, Y.B.; Carlson, B.E.; Lai, X.M. Study of resistance spot weldability of a new stainless steel. *Sci. Technol. Weld. Join.* **2019**, *24*, 101–111. [CrossRef]
7. Banga, H.K.; Kalra, P.; Kumar, R.; Singh, S.; Pruncu, C.I. Optimization of the cycle time of robotics resistance spot welding for automotive applications. *J. Adv. Manuf. Process.* **2021**, *3*, e10084. [CrossRef]
8. Pouranvari, M.; Alizadeh-Sh, M.; Marashi, S.P.H. Welding metallurgy of stainless steels during resistance spot welding part I: Fusion zone. *Sci. Technol. Weld. Join.* **2015**, *20*, 502–511. [CrossRef]
9. Akkas, N. Welding time effect on tensile-shear loading in resistance spot welding of SPA-H weathering steel sheets used in railway vehicles. *Acta Phys. Pol. A* **2017**, *131*, 52–54. [CrossRef]
10. Zhou, K.; Yao, P. Overview of recent advances of process analysis and quality control in resistance spot welding. *Mech. Syst. Signal Process.* **2019**, *124*, 170–198. [CrossRef]
11. Pouranvari, M.; Marashi, S.P.H. Critical review of automotive steels spot welding: Process, structure and properties. *Sci. Technol. Weld. Join.* **2013**, *18*, 361–403. [CrossRef]
12. Al-Mukhtar, A. Review of Resistance Spot Welding Sheets: Processes and Failure Mode. *Adv. Eng. Forum* **2016**, *17*, 31–57. [CrossRef]
13. Jou, M. Real time monitoring weld quality of resistance spot welding for the fabrication of sheet metal assemblies. *J. Mater. Process. Technol.* **2003**, *132*, 102–113. [CrossRef]
14. Wang, H.X.; Wang, C.S.; Shi, C.Y.; Xiao, J.F. The Investigation of Partial Penetration Lap Laser Welding Applied on Stainless Steel Railway Vehicles. *Adv. Mater. Res.* **2010**, *97*, 3832–3835. [CrossRef]

15. Cheon, J.Y.; Vijayan, V.; Murgun, S.; Do Park, Y.; Kim, J.H.; Yu, J.Y.; Ji, C. Optimization of pulsed current in resistance spot welding of Zn-coated hot-stamped boron steels. *J. Mech. Sci. Technol.* **2019**, *33*, 1615–1621. [CrossRef]
16. He, L.; DiGiovanni, C.; Han, X.; Mehling, C.; Wintjes, E.; Biro, E.; Zhou, N.Y. Suppression of liquid metal embrittlement in resistance spot welding of TRIP steel. *Sci. Technol. Weld. Join.* **2019**, *24*, 579–586. [CrossRef]
17. Song, S.; Shojaee, M.; Midawi, A.R.H.; Sherepenko, O.; Ghassemi-Armaki, H.; Biro, E. Influence of expulsion and heat extraction resulting from changes to electrode force on liquid metal embrittlement during resistance spot welding. *J. Mater. Res. Technol.* **2023**, *23*, 1458–1470. [CrossRef]
18. Fukumoto, S.; Fujiwara, K.; Toji, S.; Yamamoto, A. Small-scale resistance spot welding of austenitic stainless steels. *Mater. Sci. Eng. A Struct.* **2008**, *492*, 243–249. [CrossRef]
19. Jagadeesha, T. Experimental studies in weld nugget strength of resistance spot-welded 316L austenitic stainless steel sheet. *Int Int. J. Adv. Manuf. Technol.* **2017**, *93*, 505–513. [CrossRef]
20. Biradar, A.K.; Dabade, B.M. Optimization of resistance spot welding process parameters in dissimilar joint of MS and ASS 304 sheets. *Mater. Today Proc.* **2020**, *26*, 1284–1288. [CrossRef]
21. Sreehari, M.; Bhaskar, G.B. Experimental investigations on resistance spot welding for producing indentation free joints on AISI 409M grade stainless steels. *Mater. Res. Express* **2019**, *6*, 46527. [CrossRef]
22. Gong, Z.; Zhang, T.; Chen, Y.; Lu, J.; Ding, X.; Zhang, S.; Lan, M.; Shen, Y.; Wang, S. Effect of laser shock peening on stress corrosion cracking of TC4/2A14 dissimilar metal friction stir welding joints. *J. Mater. Res. Technol.* **2024**, *30*, 1716–1725. [CrossRef]
23. Han, X.; Wu, W.; Xiao, L. Effect of Spot Weld Indentation on Spot Weld Nugget Characterization. In Proceedings of the 2018 IEEE International Ultrasonics Symposium (IUS), Kobe, Japan, 22–25 October 2018. [CrossRef]
24. Pouranvari, M. Influence of welding parameters on peak load and energy absorption of dissimilar resistance spot welds of DP600 and AISI 1008 steels. *Can. Metall. Q.* **2011**, *50*, 381–388. [CrossRef]
25. Rajesh, D.; Chandran, V.; Lenin, N.; Subramanian, A.; Deva, M.; Balamurugan, P. Optimization of dual pulse resistance welding parameters for ASTM A240 stainless steel sheets: A multi-objective approach. *Interactions* **2024**, *245*, 140. [CrossRef]
26. Qi, L.; Li, F.; Zhang, Q.; Xu, Y.; Han, X.; Li, Y. Improvement of single-sided resistance spot welding of austenitic stainless steel using radial magnetic field. *ASME J. Manuf. Sci. Eng.* **2021**, *143*, 031004. [CrossRef]
27. Xia, Y.-J.; Su, Z.-W.; Lou, M.; Li, Y.-B.; Carlson, B.E. Online Precision Measurement of Weld Indentation in Resistance Spot Welding Using Servo Gun. *IEEE Trans. Instrum. Meas.* **2020**, *69*, 4465–4475. [CrossRef]
28. Xie, J.; Chen, Y.; Wang, H.; Zhang, T.; Zheng, M.; Wang, S.; Yin, L.; Shen, J.; Oliveira, J.P. Phase transformation mechanisms of NiTi shape memory alloy during electromagnetic pulse welding of Al/NiTi dissimilar joints. *Mater. Sci. Eng. A* **2024**, *893*, 146119. [CrossRef]
29. Majlinger, K.; Katula, L.T.; Varbai, B. Prediction of the Shear Tension Strength of Resistance Spot Welded Thin Steel Sheets from High- to Ultrahigh Strength Range. *Period. Polytech. Mech.* **2022**, *66*, 67–82. [CrossRef]
30. Chen, L.; Zhang, Y.; Xue, X.; Wang, B.; Yang, J.; Zhang, Z.; Tyrer, N.; Barber, G.C. Investigation on shearing strength of resistance spot-welded joints of dissimilar steel plates with varying welding current and time. *J. Mater. Res. Technol.* **2022**, *16*, 1021–1028. [CrossRef]
31. Shawon, M.R.A.; Gulshan, F.; Kurny, A.S.W. Effect of Welding Current on the Structure and Properties of Resistance Spot Welded Dissimilar (Austenitic Stainless Steel and Low Carbon Steel) Metal Joints. *J. Inst. Eng. Ser. D* **2015**, *96*, 29–36. [CrossRef]
32. Pouranvari, M.; Marashi, S.P.H. Failure Behavior of Three-Steel Sheets Resistance Spot Welds: Effect of Joint Design. *J. Mater. Eng. Perform.* **2012**, *21*, 1669–1675. [CrossRef]

metals
MDPI

Article

Effect of Bismuth and Telluride on the Inclusions of Sulfur Free-Cutting Steel

Xin Wang [1], Hongmei Zhang [1,*], Jianling Wang [2], Rui Zhu [1], Yuchuan Zhu [1], Fenglin Lu [1], Jinmeng Li [1] and Zhengyi Jiang [3,*]

1 School of Materials and Metallurgy, University of Science and Technology Liaoning, Anshan 114051, China
2 Ansteel Group Chaoyang Steel & Iron Co., Ltd., Chaoyang 122000, China
3 School of Mechanical, Materials, Mechatronic and Biomedical Engineering, University of Wollongong, Wollongong, NSW 2522, Australia
* Correspondence: zhanghm@ustl.edu.cn (H.Z.); jiang@uow.edu.au (Z.J.); Tel.: +86-138-0492-7151 (H.Z.); +86-188-1650-7540 (Z.J.)

Abstract: The development of free-cutting steel is inseparable from the development of environmentally friendly alloy elements and the control of inclusions shape. Alloying elements can affect the composition, morphology, size, and distribution of inclusion, which are the main factors affecting the machinability of free-cutting steel. This study selected sulfur free-cutting steel with different chemical compositions as the research object to examine the effects of bismuth and bismuth tellurium on sulfur-containing free-cutting steel through electrolytic corrosion experiments, metallographic microscopy, scanning electron microscopy, energy spectrum, and electron backscattering analyzer. The results showed that the microstructures of free-cutting-steel containing sulfur, free-cutting steel containing sulfur bismuth, and free-cutting steel containing sulfur bismuth tellurium are composed of ferrite, pearlite, and inclusions. The inclusions in sulfur-containing free-cutting steel are chain, cluster, and a few dotted MnS. The inclusions in sulfur-bismuth free-cutting steel are point and a few dotted MnS. After the addition of Te, the number of dotted inclusions is reduced, while the number of chain and cluster inclusions is increased. Most of the inclusions in bismuth-containing free-cutting steel are flake inclusions, and the class II MnS change into class III MnS, which is beneficial for improving the free-cutting property of steel and to reduce anisotropy. With the addition of Te, MnS of other shapes, such as heart, water drop, butterfly, etc. of a length–width ratio of less than 4 also appeared as MnS and MnTe complex inclusions, and the fusiform manganese sulfide accounted for most of the steel. Both Bi and Te had modification effects on MnS.

Keywords: free-cutting steel; inclusion; 3D etching; bismuth; tellurium

1. Introduction

Free-cutting steel can reduce processing costs, improve tool life, and reduce cutting force. It also has good economic benefits and contributes to rapid development by virtue of its excellent cutting performance in instruments, precision instruments, automobile daily necessities, and other industries in which it has been widely used. As such, its demand is rising year by year [1,2]. Its types have gradually increased to more than a dozen, among which the lead type of free-cutting steel is harmful to the environment. While ROHS, ELV, and WEEE directives all clearly specify the use of lead, lead causes harm to people's health [3–5]. At present, the development trend of free-cutting steel is "de-leading" [6]. Some researchers use low-sulfur free-cutting steel instead of lead-free free-cutting steel [7]. As a new type of easy cutting steel element, Bi has a melting point of only 271.5 °C and a boiling point of 1564 °C. It has stable chemical properties and is similar to Pb in physical properties, but it is non-toxic and will not produce toxic steam. It has soft properties and a lower density than Pb to reduce segregation in steel. They have the same machinability, they are fine metal particle inclusions uniformly distributed around sulfide, and they are affordable [8,9]. Bi in steel can prevent MnS deformation during the forging process [10]. The addition of Te makes MnS inclusions more resistant to deformation and more spherical in the hot rolling process [11]. Adding Te can improve the cutting performance of steel. Te and S cannot synthesize compounds, but the molten state can be mutually soluble. Tellurium is a highly effective and anisotropic compound that can be added to steel alone, or it can be combined with S, Se, and other free-cutting elements to form complex inclusions in steel through the interaction of different elements [12]. Compound addition of appropriate amounts of different free-cutting elements is expected to multiply the free-cutting properties of steel [13].

At present, the trends of free-cutting steel are as follows: improving the standard number of free-cutting steel; improving the quality and accuracy of the original free-cutting steel, increasing the quantity of free-cutting steel; optimizing the control technique of inclusions; exploring new environmentally friendly free-cutting elements; and developing new environmentally friendly free-cutting steels according to the principle of "multi-element and small amount" through which multi-element composite free-cutting steel is developed [14,15]. By changing the chemical composition of free-cutting steel by adding one or more free-cutting elements, especially by controlling the quantity, shape, distribution, size, and appearance of inclusions, the machinability of free-cutting steel can be improved, which is of great significance to the research of free-cutting steel [16].

Steel mills such as Inland Steel of the United States, Aichi of Japan, and Posco of South Korea have developed environmentally friendly free-cutting steel that uses bismuth instead of lead. The inclusions in Japanese steel are mostly spindle-shaped and evenly distributed. [17]. The reserves of Bi in China hold the first place in the world, and Bi can be added to replace the Pb system's free-cutting steel. Many Chinese steel mills, such as Xingcheng Special Steel, Shigang Steel, Yongsteel, Qingsteel, Hangang Steel, Guigang Steel, Shaosteel, and Angang Steel have conducted research on bismuth-containing free-cutting steel and have made many achievements. However, so far, low-sulfur bismuth-containing free-cutting steel has not been widely used for commercial purposes [18].

The machinability of free-cutting steel depends on the formation of sulfide inclusions in the steel. During processing, sulfide inclusions induce stress concentration and facilitate chip fracture. Non-metallic inclusions exist in steel in the form of second-phase particles, which have a great influence on the mechanical properties and machinability of steel [19]. Inclusions in steel are unavoidable and change the mechanical properties and machinability of steel. As a source of stress concentration, sulfide can reduce cutting force and form a protective layer in the tool [20,21].

Manganese sulfide as a stress concentration source, cut off from the metal connection, also plays the role of solid lubrication and wrapping hard points, so that the chip is easy to break, prolonging tool life and reducing cutting cost [22,23]. MnS also plays an active role in the nucleation and refinement of ferrite in grains, and the coating of harmful precipitates on the surface of MnS can improve the cutting performance and strength of steel [24]. The quality evaluation of inclusions includes the chemical composition, macroscopic and microscopic structure, characteristics, and shape of inclusions [25]. To study manganese sulfide inclusion, it is necessary to control the morphology, size, and distribution of manganese sulfide inclusion so as to avoid the production of a large inclusion, which will seriously harm the quality of steel. The best form of manganese sulfide is a small and evenly distributed spherical or fusiform inclusion. Manganese sulfide will affect the anisotropy of steel, so it is necessary to study its aspect ratio (length/width). Sims CE classified the two-dimensional morphology of manganese sulfide in steel into the following categories: class I includes spherical, fusiform, and diffuse distribution in steel formed by monotectic reaction and arbitrary distribution. The second type includes strip, rod, and fan, which are generally dispersed near grain boundaries and formed by eutectic reactions. The third group is distributed in a massive and irregular manner and is formed by pseudo-eutectic reactions [26,27]. Studies have shown that class I and class III MnS in an as-cast state are more conducive to improving cutting performance. Class II MnS have a greater impact on the anisotropy of steel and are not conducive to cutting, so the generation of class II MnS should be avoided [28]. When the MnS aspect ratio is less than three, it has better easy cutting performance; when the MnS aspect ratio is more than four, the anisotropy of the steel will also increase, and the chips will easily stick, reducing the surface machining quality [29].

In this paper, Y08 steel was used as the base steel, and the effect of chemical composition on the microstructure and inclusions of the bismuth-containing easy cutting steel was studied by adding the environmentally friendly elements bismuth and bismuth tellurium compound as the chemical composition of the bismuth-containing easy cutting steel. The inclusion was exposed through the corrosion of steel matrix. The inclusion with three-dimensional morphology can be observed under the scanning electron microscope. Compared with the common observation of two-dimensional inclusion, three-dimensional inclusion is more three-dimensional and more intuitive, which greatly helps the analysis of the morphology, distribution, and size of inclusion. The inclusions we expect to obtain are fusiform inclusions with the aspect ratio of $1 < \text{length/width} \leq 3$. The spherical and fusiform inclusions can not only improve the cutting performance of the material, but also weaken the influence on the mechanical properties of steel such as transverse impact properties. They are not easy to deform during heat treatment and reduce the anisotropy.

In this paper, the distribution, size, morphology, and composition of the inclusions were analyzed by metallographic microscope and scanning electron microscope combined with 3D saprography technology, and the effect of bismuth and bismuth tellurium complex on the inclusion of sulfur-containing free-cutting steel was studied.

2. Materials and Methods

2.1. Experimental Materials

Table 1 shows the main chemical compositions of three groups of free-cutting steels with the addition of free-cutting elements S, Bi, and Te, among which 1# contains S without Bi, 2# contains S and Bi, and 3# contains S, Bi, and Te. The first and second groups were compared without bismuth, and the second and third groups were compared by adding Te element to make multicomponent bismuth-containing free-cutting steel. The austenite grain growth behavior of the free-cutting steel after adding Te element was studied.

Table 1. Chemical composition of 3 kinds of test steels (mass fraction, %).

No.	C	Si	P	S	Mn	Bi	Te
Sample 1	0.058	0.060	0.950	0.173	0.948	-	-
Sample 2	0.043	0.020	0.800	0.088	0.796	0.004	-
Sample 3	0.078	0.070	0.870	0.131	0.874	0.011	0.025

2.2. Chemical Composition Design

Based on the composition and properties of GB/T8731-2008 steel, Y08 steel was selected as the master steel in this study, and the chemical composition was designed by referring to other papers. At the same time, considering the need to replace other machinable elements with bismuth, an environmentally friendly element, to improve its machinable properties without affecting other mechanical properties, three groups of steel were designed on this basis. The contents are shown in Table 2.

Table 2. Chemical composition design table of 3 groups of steel numbers, %.

No.	C	Si	P	S	Mn	Bi	Te
Sample 1	0.06–0.08	0.05–0.15	≤0.015	0.3–0.35	1.0–1.05	-	-
Sample 2	0.06–0.08	0.05–0.15	≤0.015	0.15–0.18	1.0–1.05	0.05	-
Sample 3	0.06–0.08	0.05–0.15	≤0.015	0.25–0.35	1.0–1.05	0.05	0.05

The first group: As the control group, according to the national standard Y08 and considering the improvement of steel's free-cutting performance, the sulfur content was set at 0.3~0.35%. C can be a solid solution for strengthening and is an important element to improve strength, reduce carbon content (which will increase debris nodular), hinder tool wear, and improve the material's easy cutting performance. As such, the design's carbon content was 0.06~0.08%. P can be solidly soluble in ferrite to improve the machining surface finish; design P content was less than 0.015%.

The second group: On the basis of group 1, the contents of C, Si, Mn, and P remained unchanged, and Bi was added. Due to the addition of Bi element, the content of S element was reduced by 1/2 relative to the first group to create bismuth-containing sulfur.

The third group: On the basis of the second group, the contents of C, Si, Mn, P, and Bi were unchanged, and the content of sulfur was also increased after the addition of Te element. The design content was 0.25~0.35% to make the sulfur, bismuth, and tellurium ternary complex.

The carbon, silicon, manganese, phosphorus, and sulfur contents of the three groups of steels were the same. Steel 1# was designed according to the national standard Y08, and the sulfur content was controlled within 0.3~0.35%, so that the machinability of the steel could be improved under the condition of high sulfur and no bismuth. The microstructure and inclusion state of steel 1# designed in this composition were compared with those of the 2# and 3# steels containing other easily machinable steels. The effect of bismuth element on the microstructure and inclusions of free-cutting steel was studied by comparing 2# with 1#. The effect of bismuth–tellurium complex on the microstructure and inclusions of bismuth-containing easy cutting steel was studied by adding the bismuth–tellurium complex elements and selecting the appropriate sulfur content.

2.3. Metallographic Test

The three test steels needed to be ground, polished, and etched before being examined under a metallographic microscope. The test steel was machined into a size of 5 mm × 10 mm × 20 mm, and the 10 mm × 20 mm surface of the test steel was polished. The samples were polished successively on W28 sandpaper (320 mesh), W20 sandpaper (500 mesh), W14 sandpaper (600 mesh), and W10 sandpaper (800 mesh) until the surface scratches were all in the same direction. Then, the polishing machine was used to polish the samples. After polishing, the surface was cleaned with an aqueous ethanol immediately, any dirt was removed, and the surface was dried with a hair dryer. In order to observe the tissue inside the sample, the surface of the sample was wiped and etched with cotton dipped in 4% volume concentration of nitrate ethanol solution, the surface was rinsed with an aqueous ethanol, and then the tissue was observed with an ultra-depth-of-field microscope. An ultrasonic cleaning machine was used to clean the pattern and then it was dried with a hair dryer. The microscope used in this study was a VHX-5000 metallographic microscope.

2.4. 3D etching Technology

In order to better observe the morphology of inclusions, corrosion of steel matrix by electrolytic etching technology to retain inclusions was conducive to visually seeing the three-dimensional morphology of inclusions. Before electrolytic etching testing the steel, the experimental steel was processed into 5 × 10 × 20 mm. The experimental conditions were as follows: electrolyte (70% methanol, 20% acetyl acetone, and 10% tetramethylammonium chloride), current (0.449 A), electrolytic time (1 h), electrolytic temperature (−1~10 °C), and electrolytic pH (neutral or weakly acidic). Figure 1 shows the electrolytic power supply and electrolytic reagent (methanol, acetyl acetone, and tetramethylammonium chloride)

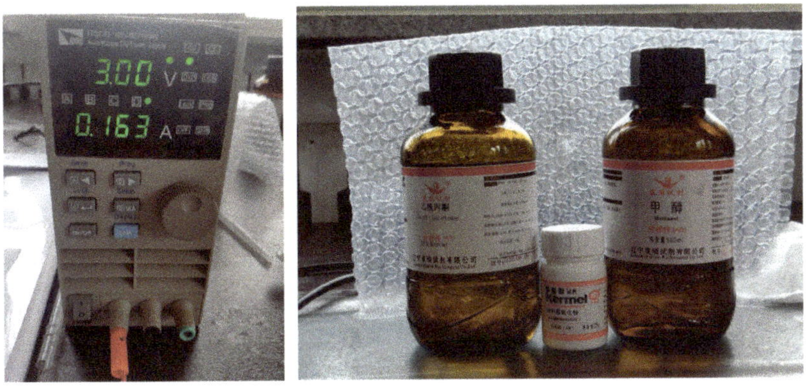

Figure 1. Electrolytic power supply (IT6702) and electrolytic reagent.

In these conditions, the electrolytic device, as shown in Figure 2, was used for the electrolytic experiment. Electrolytic test equipment included: (1) water bath device, (2) stainless steel, (3) protective cover, (4) thermometer, (5) test steel, (6) electrolyte, and (7) power supply. The cathode was connected with the stainless steel and the anode was connected with the experimental steel. Except for the electrolytic polishing surface, the remaining surface of the test steel was wrapped in paraffin wax to prevent other surfaces from being corroded. The polished electrolytic sample was wiped with alcohol and put it into the electrolytic device. Attention was paid to the electrolytic surface facing the stainless steel sheet. Then, the power supply was adjusted to the appropriate current size, and the time and temperature were well-controlled to conduct the experiment. After electrolytic corrosion of the test steel, it was necessary to clean it with anhydrous ethanol, and the morphology of inclusions in the three samples was observed by scanning electron microscopy.

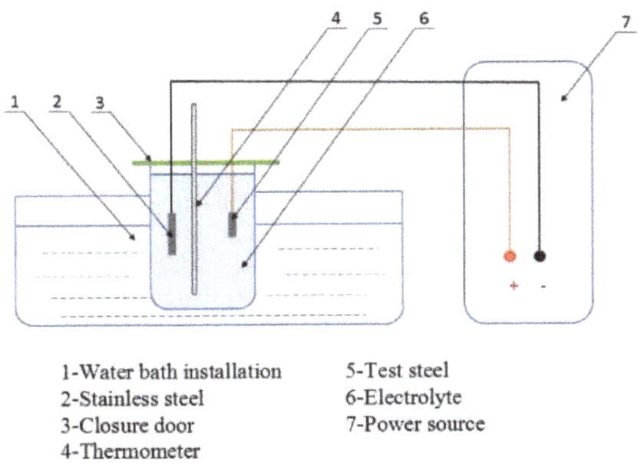

Figure 2. Electrolysis plant.

3. Results and Discussion

3.1. Effect of Chemical Composition on Quantity and Distribution of Inclusions

The three groups of steels after forging under metallographic microscope are shown in Figure 3, in which the inclusion is in the red frame, the ferrite is in the green frame, and the pearlite is in the blue frame. Figure 3a shows the microstructure of 1#, which is mainly composed of pearlite, ferrite, and inclusions. The pearlite is relatively large in size and distributed in chain shape, while the inclusions in steel are distributed in chain shape, cluster shape, less spot shape, and in large quantity. Figure 3b shows the microstructure of 2#, which is mainly composed of pearlite, ferrite, and inclusions. The pearlite is small in size, and the amount is the highest of the three groups of steels. The inclusions are dotted and less chained in the steel matrix. Figure 3c shows the microstructure of 3#, which is mainly composed of pearlite, ferrite, and inclusions. The inclusions are distributed in clusters, chains, and a few dots.

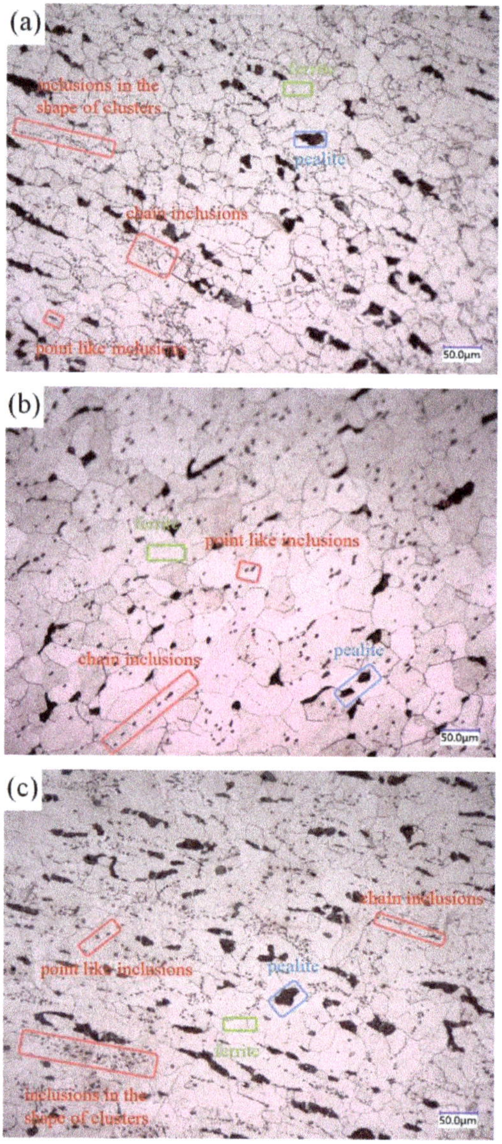

Figure 3. Metallographic pictures of 3 groups of steel: (**a**) 1#; (**b**) 2#; (**c**) 3#.

As can be seen from Figure 3, the inclusions in sulfur-type free-cutting steel are mainly distributed in chain shape, the addition of Bi reduces the number of chain inclusions, the number of dot inclusions increases, and the distribution of inclusions is uniform. After the compound addition of Bi and Te, the number of cluster inclusions increases, among which the free-cutting steel (in which the environmental protection type of free-cutting element Bi is added alone) has the best distribution.

According to the carbon content of the three groups of steels, the low temperature microstructure is proeutectoid ferrite and pearlite. The ratio of ferrite to pearlite of the three groups of steels can be calculated by using the lever theorem, and the formula is as follows:

$$\text{Ferrite} = \frac{0.77 - \text{w(c)}}{0.77 - 0.0218} \times 100\% \quad (1)$$

$$\text{Pearlite} = \frac{\text{w(c)} - 0.0218}{0.77 - 0.0218} \times 100\% \quad (2)$$

The proportion of ferrite and pearlite of 1# steel is 95.2% and 4.8% by calculation. The proportion of ferrite and pearlite of 2# steel is 97.2% and 2.8%, respectively. Finally, 3# steel has 92.5% ferrite and 7.5% pearlite. Again, green and blue are indeed proeutectoid ferrites and pearlite, according to the proportions of colors in the figure.

3.2. Observation and Analysis of Inclusions by Scanning Electron Microscopy

Three groups of forged steel were observed by scanning electron microscopy. Among them, Figure 4 is the SEM image of 1# under the field of view of 500 times. It is found that the inclusions are spherical, fusiform, and individual rod-like, and the inclusions are distributed in chain shape. Statistical analysis of inclusions in the image shows that the total number of inclusions is 186, the maximum size is 8.63 μm, the minimum size is 0.81 μm, and the average size is 2.60 μm, among which the number of inclusions is the largest between 1.8~2.7 μm, and the proportion is close to 40% of the total sample value.

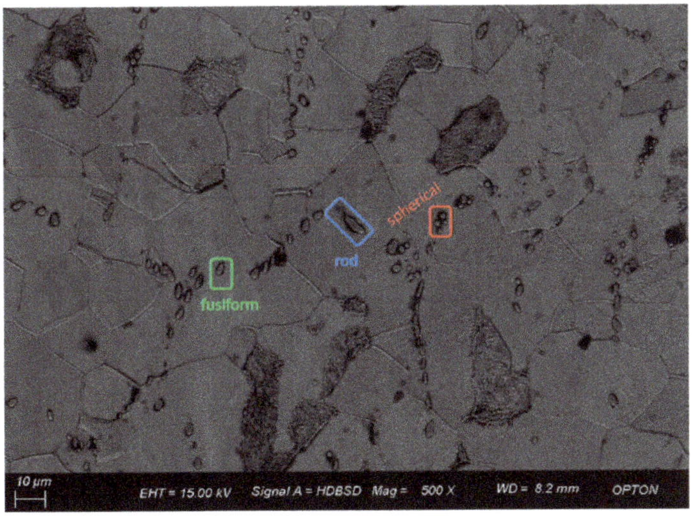

Figure 4. Morphology and quantity distribution of No. 1 steel inclusions.

Figure 5 shows the SEM image of 2# under a field of view of 500 times. It is found that the inclusions are spherical, fusiform, and rod-shaped, among which the shapes are spherical and fusiform, accounting for nearly 40% of the total sample amount. Statistical analysis of inclusions in the image shows that the number of inclusions is 59, the maximum size is 18.84 μm, the minimum size is 0.76 μm, the average grain size is 4.43 μm, and the size is less than 1.9 μm.

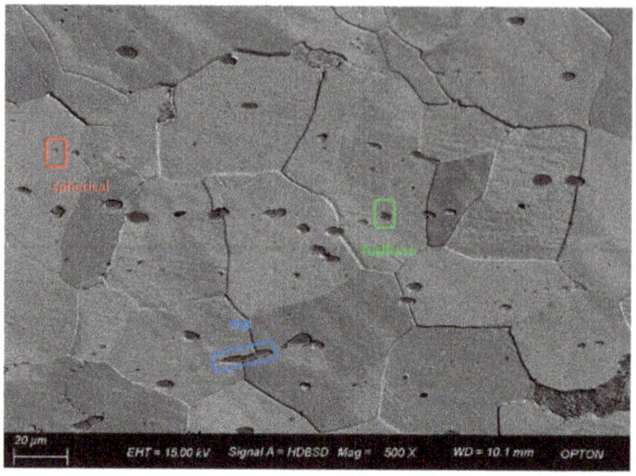

Figure 5. Morphology and quantity distribution of No. 2 steel inclusions.

Figure 6 is the SEM image of 3# under a field of view of 500 times. Through observation of the microscopic structure and inclusions of 3#, it is found that the inclusions have various shapes, mainly spherical, spindle, and rod-shaped, among which the large inclusions have a small number and are generally rod-shaped, while the inclusions are distributed in chain shape. The statistical analysis of inclusions in the image shows that the number of inclusions is 147, the maximum size is 13.56 μm, the minimum size is 0.63 μm, and the average grain size is 2.48 μm. When the size is between 1.4 and 2.8 μm, the number of samples is the maximum, which is close to 50% of the total sample size.

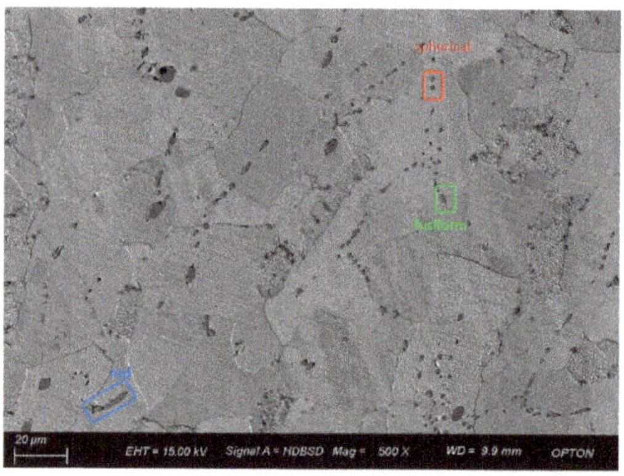

Figure 6. Morphology and quantity distribution of No. 3 steel inclusions.

3.3. Influence of Chemical Composition on the Morphology, Size, and Shape of Inclusions

Figure 7 shows the EBSD diagram of the three groups of steel after electrolysis. As shown in Figure 7a, 1# includes type I and type II inclusions. Spherical inclusions with a size of about 2 μm, fusiform inclusions with a size of 3–8 μm, and rod–rod inclusions with a size of 8–14 μm primarily exist in the steel matrix and their distribution is uneven; the length/width of spherical inclusions is close to 1. The length/width of fusiform inclusions is between 1.5 and 2.1, and that of rod–rod inclusions is above 3.7. The size of the largest rod–rod inclusions is 14.6 μm and length/width is 4.98. As can be seen in Figure 7b, 2# includes type I and type III inclusions, mainly composed of irregular flake and inclusions over 10 μm, a small number of 2 μm spherical inclusions and about 4 μm fusiform inclusions, and large flake inclusions with length/width ranges from 2.1 to 2.9; the longest inclusions are 17.7 μm long and length/width is 2.2. Although the size of the flake inclusion is large, its width is also large, so the ratio of aspect to width is relatively low. As shown in Figure 7c, 3# includes type I, type II, and type III inclusions of different shapes, including fusiform, spherical, heart-shaped, droplet, butterfly, and irregular, short rod-shaped inclusions. The length of rod-shaped inclusions with the largest individual size reaches 19.66 μm and length/width is 5.1, while the length of other rod-shaped inclusions is 9–13 μm and length/width is about 3.7.

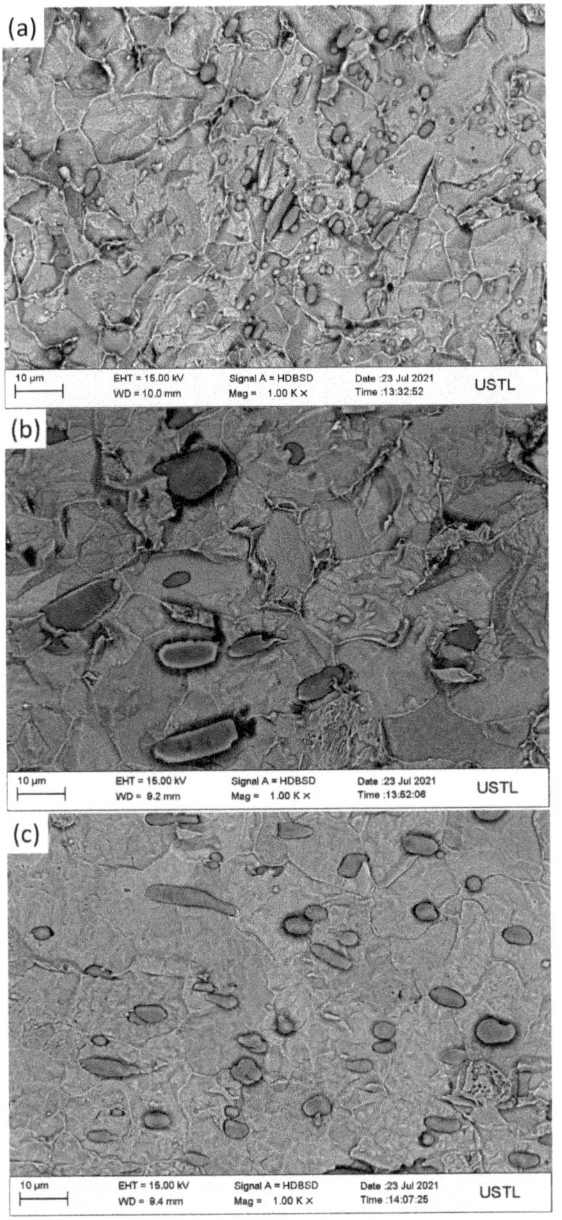

Figure 7. Morphologies observation of inclusions in three groups of steels after corrosion: (**a**) 1#; (**b**) 2#; (**c**) 3#.

The inclusions of sulfur-type steel are spherical, fusiform, and a few rod-like, with a size of 2~14.6 μm. After the addition of Bi, the total number of inclusions is reduced, and the flake inclusions are increased, with a size of 2~17.7 μm. After the addition of Bi and Te, the number of inclusions is less in 1# than 2#, the number of rod-like inclusions is less, and the morphology of inclusions is changed. Inclusions of other shapes appeared, ranging in size from 2 to 19.66 μm. The number of inclusions in 2# steel is the least among the three groups. The addition of Bi alone and Bi and Te compound can improve the shape of inclusions and reduce the aspect ratio of inclusions. However, some inclusions still have large sizes.

3.4. Influence of Chemical Composition on the Morphology, Size, and Shape of Inclusions

Figure 8 shows the EDS line sweep of 1# after electrolysis, and Figure 8a shows the spindle-shaped and rod-shaped inclusions, which can be judged as MnS by energy spectrum analysis. Figure 8b shows the fusiform and spherical inclusions of line sweep, which are also MnS, as found by energy spectrum analysis.

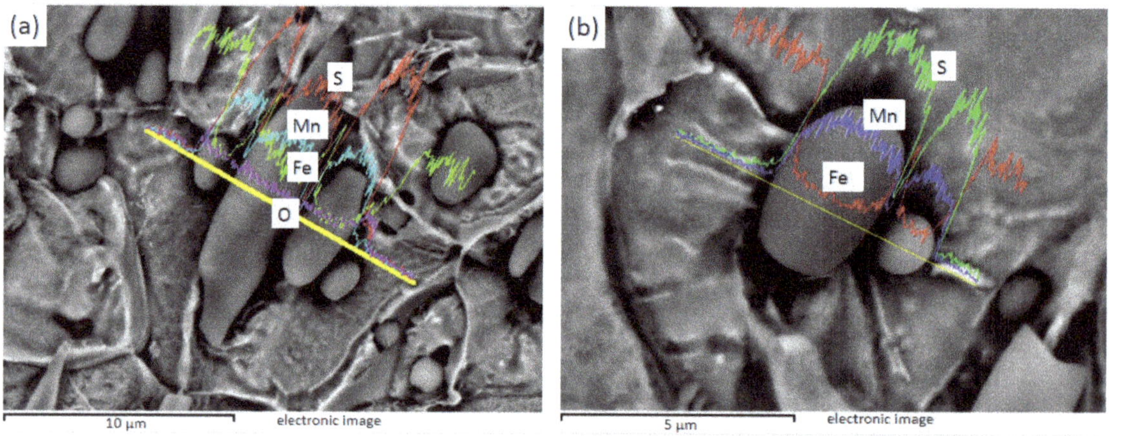

Figure 8. 1# Spherical, fusiform, and rod-bar MnS inclusions: (**a**) spherical, fusiform, and rod-bar MnS; (**b**) spherical and fusiform MnS.

Figure 9 shows the EDS line sweep of 2# after electrolysis, and Figure 9a shows the plat-like inclusion, which can be judged as MnS by energy spectrum analysis. Figure 9b is a crescent-shaped MnS inclusion, and Figure 9c is a spherical MnS inclusion.

Figure 10 shows the EDS line sweep of 3# after electrolysis. Figure 10a shows the MnS inside the Te package, and Bi is distributed in the MnS. Figure 10b shows the heart-shaped MnS, and Figure 10c shows the MnS with unilateral Te. It can be found that Te exists in steel mainly wrapped in one or both sides of MnS inclusions, while Bi can exist inside MnS.

The inclusions in sulfur-containing free-cutting steel are MnS, the inclusions in sulfur-containing bismuth free-cutting steel are MnS, Mn (Te, S), and Mn (Te, S), and Bi composite inclusions are found in sulfur-containing bismuth tellurium free-cutting steel. After Bi element was added, the size of inclusions tended to increase, the total number of inclusions decreased, the rod-bar inclusions decreased, and the aspect ratio of inclusions decreased obviously. The addition of Bi and Te changed the composition and morphology of the inclusions, and the shape of the inclusions was more diverse. Bi and Te have obvious improvement effects on inclusions, as both can reduce the number of inclusions and reduce the rod-like inclusions with large aspect ratio.

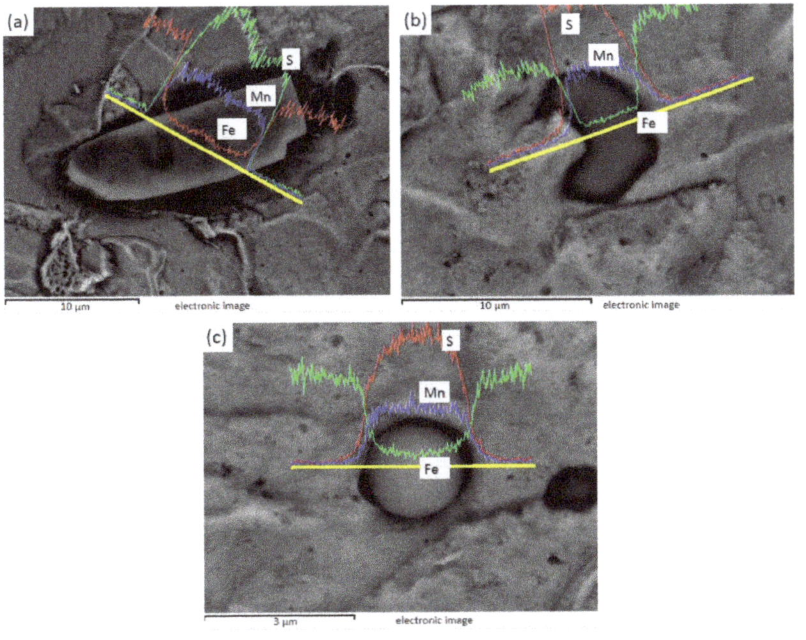

Figure 9. 2# Plate, crescent, and globular MnS inclusions: (**a**) plate MnS; (**b**) crescent MnS; (**c**) globular MnS.

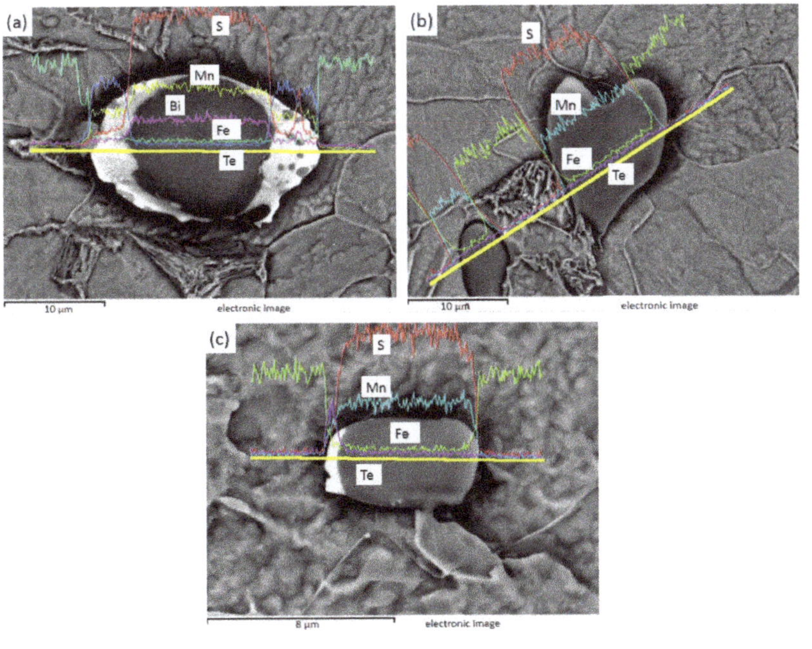

Figure 10. 3# MnS inclusions of heart shape and with Te on either side or one side: (**a**) fusiform MnS containing Te on both sides and Bi in the middle; (**b**) heart-shaped MnS; (**c**) MnS with Te on one side.

4. Conclusions

In order to study the influence of Bi element and Bi and Te composite on the inclusions in free-cutting steel, the element design was carried out and the number, distribution, composition, size, and 3D morphology of inclusions were observed. The following conclusions were drawn:

- From the perspective of organization, the microstructure of free-cutting steel is mainly composed of ferrite, pearlite, and inclusions. The highest ferrite content is 97.2% and the lowest pearlite content is in the bismuth-containing free-cutting steel with the mass fraction of 0.088% sulfur and 0.004% Bi. The content of ferrite in bismuth-containing free-cutting steel with 0.131% sulfur, 0.011% Bi, and 0.025% Te is the lowest (92.5%) and the highest pearlite is 7.5%.
- In terms of quantity and distribution, the inclusion quantity of bismuth-containing free-cutting steel with sulfur content of 0.088% and Bi content of 0.004% is the least, and the inclusion quantity of sulfur-containing free-cutting steel with sulfur content of 0.173% is the most. The inclusions in the three groups of free-cutting steels are dotted, chained, and clustered. After the addition of Bi and Te, the distribution of cluster inclusions increases, while after the addition of Bi, the point inclusions increase and are evenly distributed, which is the optimal distribution pattern of free-cutting steel.
- From the perspective of morphology and size, the inclusions of 0.173% sulfur-containing free-cutting steel are spherical, fusiform, and rod-like, with sizes ranging from 2 to 14.6 μm. Some rod-like inclusions have an aspect ratio of 3.7, which will affect the anisotropy of the steel. The inclusion morphology of bismuth-containing free-cutting steel with sulfur content of 0.088% and Bi content of 0.004% is mainly flake inclusion. The overall size is larger, ranging from 13 to 17.7 μm, and the aspect ratio is 2.1 to 2.9, which is also an ideal class III inclusion. The inclusions of bismuth-containing tellurium in free-cutting steels with 0.131% sulfur, 0.011% Bi, and 0.025% Te have different shapes and sizes ranging from 2 to 19.66 μm, but the second type of rod-like inclusions still exist. After adding Bi, the inclusions tend to transform to type III and the size of inclusions increases. After adding Bi and Te, the morphology of inclusions changed irregularly, and the morphology was different.
- In terms of composition, the inclusions of sulfur-containing free-cutting steel and bismuth-containing free-cutting steel are MnS and (Mn, Fe) S, with spherical, fusiform, and rod-like shapes. The inclusions of MnS, Mn (Te, S), and Mn (Te, S) mixed with Bi in sulfur-containing bismuth tellurium steel are three kinds of inclusions, and where Te is distributed at the tip of MnS, Bi may exist in MnS.

Author Contributions: Writing—original draft preparation, X.W.; conceptualization, H.Z., J.W. and Z.J.; investigation, R.Z.; writing—review and editing, Y.Z.; methodology, J.L.; supervision, F.L. All authors have read and agreed to the published version of the manuscript.

Funding: This research was funded by the National Natural Science Foundation of China (NSFC, No. 52274338) and the Education Department Foundation of Liaoning Province (Grant No. LJKZ0303).

Data Availability Statement: Not applicable.

Conflicts of Interest: The authors declare no conflict of interest.

References

1. Croccolo, D.; De Agostinis, M.; Fini, S.; Olmi, G.; Robusto, F. Coating effect on the fatigue strength of a free cutting steel. *J. Mech. Eng. Sci.* **2019**, *233*, 095440621983644. [CrossRef]
2. Martinez Krahmer, D.; Urbicain, G.; Sánchez Egea, A.J. Dry machinability analyses between free cutting, resulfurized, and carbon steels. *Mater. Manuf. Process.* **2020**, *35*, 460–468. [CrossRef]
3. EU-Directive. *Restriction of the Use of Certain Hazardous Substances in Electrical and Electronic Equipment (RoHS)*; European Union: Brussels, Belgium, 2003.
4. European Union. Directive 2012/19/EU of the European Parliament and of the Council of 4 July 2012 on waste electrical and electronic equipment (WEEE). *Off. J. Eur. Union L* **2012**, *197*, 38–71.

5. Wang, Y.; Bao, Y. Control of inclusions in BN type free-cutting steel and its influence on cutting performance. *J. Iron Steel Res. Int.* **2017**, *29*, 382–390. (In Chinese)
6. Zhuang, L.; Di, W. Effect of Free-cutting Additives on Machining Characteristics of Austenitic Stainless Steels. *J. Mater. Sci. Technol.* **2010**, *6*, 839–844.
7. Lin, S.G.; Yang, H.H.; Su, Y.H.; Chang, K.L.; Yang, C.H.; Lin, S.K. CALPHAD-assisted morphology control of manganese sulfide inclusions in free-cutting steels. *J. Alloys Compd.* **2019**, *779*, 844–855. [CrossRef]
8. Imanami, Y.; Tomita, K.; Nishimura, K. *Factors Affecting Surface Roughness of Low Carbon Resulfurized Free Cutting Steel*; JFE Technical Report. 23; JFE Steel Corporation: Tokyo, Japan, 2018.
9. Hu, S.; Li, Z.; Fan, T.; Fu, J. Effect of Bismuth on Sulfide in High Sulfur Free-cutting Steel. *IOP Conf. Ser. Mater. Sci. Eng.* **2019**, *611*, 012017. [CrossRef]
10. Lin, T.C.; Zhu, R.; Wang, C.J.; Li, M.G. Effect of Trace Bi on Inclusion Morphology in Free-Cutting Steel. *Iron Steel* **2013**, *48*, 77–80.
11. Sathyamurthy, P.; Vetrivelmurugan, R.; Sooryaprakash, J. Improving the machinability of leaded free cutting steel through process optimization. *IOP Conf. Ser. Mater. Sci. Eng.* **2018**, *314*, 012019. [CrossRef]
12. Wang, Z.; Shao, Z.; Li, Z.; Zhang, H.; Wang, J.; Zhang, H.; Yuan, J. Making of Free Cutting Austenitic Stainless Steels with Additions of Sulfur, Rare Earth and Bismuth. In Proceedings of the International Conference on Advanced Design & Manufacturing Engineering, Shenzhen, China, 19–20 September 2015.
13. Sun, K. Experimental study on Calcium-sulfur composite free-cutting steel. *Spec. Steel* **1983**, *1*, 71–81. (In Chinese)
14. Chen, L.; Tu, S.; Chen, W.; Li, L.; Lu, Q.; Wang, Q. Low carbon high sulfur tin antimony composite free cutting steel. *J. Univ. Sci. Technol. Beijing* **2009**, *31*, 4. (In Chinese)
15. Gao, P. Research and Development of Free-Cutting Stainless Steel for New Ballpoint Pen Head. Master's Thesis, Jiangsu University, Zhenjiang, China, 2009. (In Chinese).
16. Yuan, L.; Zheng, L.G.; Hu, X.Q.; Zhang, Y.T. Effect of Rare Earth Ce on Machinability, Strength and Toughness of 42CrMo Cut Steel. *J. Shenyang Ligong Univ.* **2022**, *41*, 76–80. (In Chinese)
17. Zhang, P.; Zeng, Z.; Fan, T.; Shen, P.; Fu, J. Deep Analysis of Free-cutting Phase and Its Distribution in Japanese SF20T Pen Tip Steel. *IOP Conf. Ser. Mater. Sci. Eng.* **2019**, *611*, 012019. [CrossRef]
18. Liu, H.T.; Chen, W.Q. Hot ductility of eco-friendly low carbon resulphurised free cutting steel with bismuth. *Ironmak. Steelmak.* **2014**, *41*, 19–25. [CrossRef]
19. Zhang, Q.; Min, Y.; Xu, H.; Liu, C. Formation and Evolution of Inclusions in Si-killed Resulfurized Free-cutting Steel. *ISIJ Int.* **2018**, *58*, 1250–1256. [CrossRef]
20. Ryabov, A.V. Medium-Carbon Free-Cutting Steel. *Mater. Sci. Forum* **2019**, *946*, 47–52. [CrossRef]
21. Yan, J.; Li, T.; Shang, Z.; Guo, H. Three-dimensional characterization of MnS inclusions in steel during rolling process. *Mater. Charact.* **2019**, *158*, 109944. [CrossRef]
22. Zhang, Q. Formation and evolution behavior of inclusions in Al-killed resulfurized free-cutting steel with magnesium treatment. *J. Iron Steel Res. Int.* **2020**, *27*, 12. [CrossRef]
23. Hashimura, M.; Mizuno, A.; Miyanishi, K. Effect of MnS distribution on machinability in low-carbon free-cutting steel. *Iron Steel Technol.* **2009**, *6*, 45–51.
24. Shao, X.; Wang, X.; Jiang, M.; Wang, W.; Huang, F. Effect of Heat Treatment Conditions on Shape Control of Large-sized Elongated MnS Inclusions in Resulfurized Free-cutting Steels. *ISIJ Int.* **2011**, *51*, 1995–2001. [CrossRef]
25. Li, H.; Gao, H. Morphology of manganese sulfide in steel and its effect on steel properties. *CFHI Technol.* **2004**, *4*, 3. (In Chinese)
26. Qiu, B.; Sui, H.; Che, D.; Sun, M.; Li, J.; Wang, L. Effects of oxygen content and forging ratio on sulfide morphology and machinability in free-cutting stainless steel. *J. Iron Steel Res. Int.* **2021**, 33, 418–425. (In Chinese)
27. Zhu, Q.; Li, J.; Deng, X.; Tian, Q.; Zeng, Z.; Fu, J. Analysis of inclusions in Magnesium series easy cutting steel. *Iron Steel Vanadium Titan.* **2021**, *42*, 179–187, 192. (In Chinese)
28. Lu, J.; Cheng, G.; Qiu, W.; Long, H.; Liu, C. Effect of Zr addition on morphology characteristics of MnS inclusions in Medium carbon and high sulfur free-cutting steel. *J. Iron Steel Res. Int.* **2022**, *34*, 963–972. (In Chinese)
29. Wang, Y. Effect of forging process on mechanical properties and manganese sulfide in easy cutting steel. *Forg. Stamp. Technol.* **2021**, *46*, 12–18. (In Chinese)

metals

MDPI

Article

Dislocation Strengthening and Texture Evolution of Non-Oriented Fe-3.3 wt% Si Steel in Double Cold Rolling

Yijing Gao [1], Yunbo Xu [1,*], Haoran Chen [1], Bingyu Yuan [1], Zhenyu Gao [2] and Lifeng Zhou [3,*]

1 The State Key Laboratory of Rolling and Automation, Northeastern University, Shenyang 110819, China; 2010216@stu.neu.edu.cn (Y.G.); 2370344@stu.neu.edu.cn (H.C.); 2010217@stu.neu.edu.cn (B.Y.)
2 Silicon Steel Division of Angang Steel Co., Ltd., North of Angang Factory, Anshan 114009, China; aggzy@163.com
3 Section of Environmental Protection Key Laboratory of Eco-Industry, School of Metallurgy, Northeastern University, Shenyang 110819, China
* Correspondence: yunbo_xu@126.com (Y.X.); zhoulf@smm.neu.edu.cn (L.Z.)

Abstract: An excellent Fe-3.3 wt% Si steel was fabricated by double cold rolling and final annealing. The evolution of the microstructure and texture was studied by optical microscope (OM), X-ray diffraction (XRD), ex situ, and quasi-in situ electron backscattered diffraction (EBSD) to investigate the recrystallization behavior. Double cold rolling significantly reduced the adverse γ texture in the final annealed sheets, and a stronger η texture was observed. With a reduction ratio of 50% and 65% during double cold rolling, the γ texture almost disappeared, whereas the η texture was obviously improved. Consequently, the texture factor reached its peak, leading to a reduction in iron loss and an enhancement of magnetic induction. By combining texture regulation with dislocation strengthening, the magnetic properties of Fe-3.3 wt% Si steel were improved, and the yield strength also increased. The final sheet exhibiting exceptional magnetic characteristics and enhanced strength attained a reduction in iron loss ($P_{10/400}$ = 21.84 W/kg) of 6.43 W/kg, along with an enhancement of magnetic induction (B_{50} = 1.698 T) of 0.038 T and yield strength ($R_{p0.2}$ = 578 MPa) of 37 MPa compared to a single-stage cold rolling process.

Keywords: non-oriented silicon steel; double cold rolling; dislocation strengthening; texture evolution; quasi-in situ EBSD; microstructure

1. Introduction

Non-oriented silicon steel is a type of electrical steel containing 2–3% silicon [1,2]. Non-oriented silicon steel exhibits good permeability in all directions and uniform electromagnetic properties, making it superior to alternating current electromagnetic field performance [3]. Owing to its exceptional magnetic characteristics, non-oriented silicon steel is extensively utilized for manufacturing core materials for generators, transformers, and motors [4,5]. In contrast to conventional products, advanced non-oriented silicon steel not only maintains exceptional electromagnetic performance but also provides increased yield strength, enhancing durability and reliability [6]. Dislocation strengthening to enhance strength represents a form of resource-efficient high-strength electrical steel [7]. Zhong et al. [8] and Yu et al. [9] demonstrated that recrystallization annealing significantly improved magnetic properties but negatively impacted mechanical properties. Lin et al. [10] and Zhang et al. [11] indicated that the decline in iron loss and yield strength with increased recrystallization rate was ascribed to the substantial reduction in dislocation density during the recrystallization process. Thus, the challenge of using dislocation strengthening methods lies in enhancing the mechanical properties without compromising the magnetic performance, which remains a key research focus.

When using the traditional single-stage rolling method to produce silicon steel, a substantial reduction ratio in cold rolling can result in the development of strong γ

and α textures in the final annealed sheets [12]. Therefore, double cold rolling has been proposed to improve the recrystallization textures and optimize the magnetic properties of annealed sheets [13]. Qin et al. [14] and Yao et al. [15] demonstrated that the double cold rolling route introduced a strong η texture to the annealed sheets. Li et al. [16] indicated that a two-step rolling method remarkably weakened the detrimental γ texture while enhancing the λ fiber texture, thereby improving magnetic characteristics. Lu et al. [17] examined the impact of a two-stage rolling route on 4.5 wt% Si non-oriented electrical steel, finding that a moderate reduction can achieve excellent mechanical and magnetic properties. To enhance the regulation of the microstructure and texture and to improve the magnetic and mechanical properties, it is crucial to deepen our understanding of the mechanisms that affect the texture evolution and recrystallization behavior of high-strength silicon steel subjected to the double cold rolling process.

In recent years, characterization techniques using quasi-in situ electron backscattered diffraction (EBSD) have provided new insights into the study of material recrystallization behavior [18–20]. Quasi-in situ EBSD technology offers direct visualization of microstructure changes during the recrystallization process, making it an effective tool for in-depth analysis of texture evolution mechanisms in silicon steel [21]. In this study, Fe-3.3 wt% Si non-oriented silicon steel was produced through double cold rolling processes and dislocation strengthening. This study investigated the microstructure evolution in double cold rolling processes and the impact of dislocation strengthening on magnetic and mechanical properties. Using quasi-in situ EBSD techniques, this study elucidated the texture evolution and strengthening mechanisms of Fe-3.3 wt% Si steel during rolling and annealing, providing theoretical support for industrial production processes.

2. Materials and Methods

The experimental material was a Fe-3.3 wt% Si steel hot-rolled sheet with a thickness of 2 mm; its specific composition is listed in Table 1. Figure 1 illustrates the process flow. Four samples measuring 120 mm (RD, rolling direction) × 100 mm (TD) were cut from a hot-rolled sheet via wire cutting. Three of the samples were rolled to intermediate thicknesses of 1.2, 1.0, and 0.8 mm with multiple passes along the rolling direction (processes A, B, and C, respectively). Subsequently, the three samples were placed in a box furnace at 980 °C for 4 min, followed by air cooling to room temperature, which was further rolled to 0.35 mm. The last sample underwent direct single-stage cold rolling to 0.35 mm (process D). Finally, all specimens were annealed at 750 °C in a 100% N_2 atmosphere. Samples with different annealing processes and annealing times were named according to "process-annealing time", for example, "B-15m" indicates the sample of process B and annealing for 15 min.

Table 1. The chemical composition of the Fe-3.3 wt% Si steel (wt.%).

C	Si	Mn	Al	S	N	Fe
≤0.005	3.3	≤0.5	≤1.1	≤0.005	≤0.005	Bal.

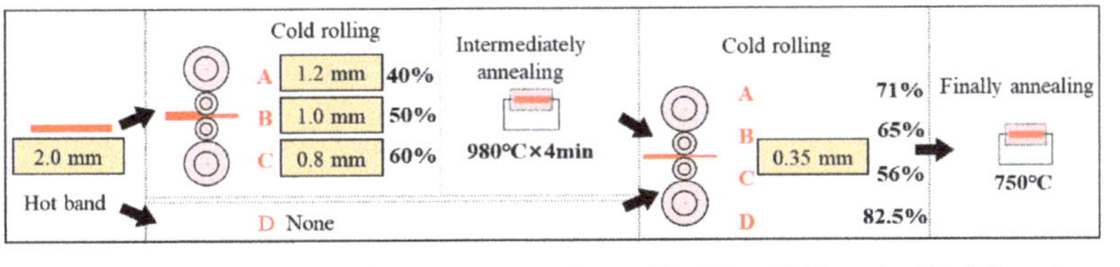

Figure 1. Schematic diagram showing process flow of Fe-3.3 wt% Si steel with different processes.

The metallographic and EBSD samples were ground using sandpapers from 400# to 2000#, followed by mechanical polishing until the surface was bright. Metallographic samples were corroded using a 4% nitric acid alcohol solution. Microstructures were observed using an optical microscope (OM). The EBSD samples used for grain orientation data collection were electrolytically polished using 10% $HClO_4$ and 90% C_2H_5OH at a polishing voltage of 25–27 V and for a polishing time of 10–15 s. For the quasi-in situ experiments, the polishing time was shortened to 3–5 s. An EBSD system was applied for grain orientation data collection. The grain sizes were calculated using at least five 100× metallographic photographs or EBSD maps and averaged. X-ray diffraction (XRD) with Co Kα diffraction was used to detect three incomplete polar patterns {100}, {110}, and {211} and calculate the Orientation Distribution Functions (ODFs). The macrotexture specimen size was 20 mm (RD) × 18 mm (TD). Standard tensile tests were adopted to measure the mechanical properties with a stretching speed of 2 mm·min^{-1}. A single-sheet magnetic detector was adopted for the measurement of magnetic properties, including magnetic flux density B_{50}, iron loss $P_{10/400}$ and $P_{15/50}$. Each group of samples was measured at least five times and averaged.

3. Results and Discussion

3.1. Microstructure and Textures of Hot-Rolled Sheet

Figure 2 shows the microstructure and textures of the initial hot-rolled sheet. As shown in Figure 2a, the microstructure was divided into a surface layer (region A), subsurface layer (region B), and central layer (region C) along the thickness direction. The surface layer (region A) exhibited a uniform and fine equiaxed microstructure. The texture was weak and dispersed, primarily consisting of {110}<001> (intensity 2.72), {881}<126> (intensity 2.76), and {661}<233> (intensity 2.66) components, as illustrated in Figure 2b. Subsurface region B was characterized by a mixed microstructure consisting of elongated deformed bands and several recrystallized grains. The texture was primarily composed of a strong Goss component and a Cube component, with the texture intensity significantly increasing compared to the surface layer, reaching a maximum of 18.06, as shown in Figure 2c. The central layer (region C) was composed of elongated deformed bands, with the texture characterized by strong α- and λ-fiber textures, as shown in Figure 2d. Both α and λ fiber textures were continuously distributed, with the λ fiber texture extending from {001}<110> to {001}<100>, exhibiting a maximum intensity of 24.59 at the Cube component. The α fiber texture extends from {001}<110> to {111}<110>, with a peak intensity of 14.46 observed at the {118} texture.

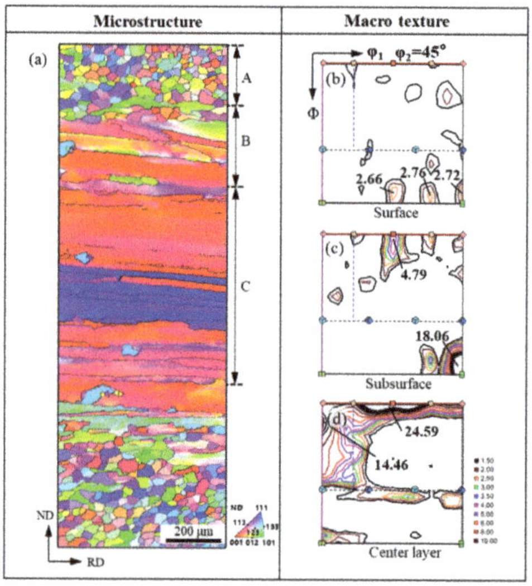

Figure 2. Microstructure and textures of initial hot-rolled band (**a**) Microstructure; (**b**) Surface texture; (**c**) Subsurface texture; (**d**) Central layer texture.

3.2. Microstructures and Textures of Cold-Rolled and Intermediate Annealed Sheets

Figure 3 shows the microstructures and textures of the first cold-rolled and intermediate annealed sheets in double cold rolling processes. The cold-rolled microstructures were similar to that of the hot-rolled sheet, exhibiting different microstructural characteristics in the direction of thickness. After the first cold rolling, the recrystallized grains on the surface were flattened and fragmented, and as the rolling reduction rate increased, the fracture degree of the equiaxed crystals also rose. In process C, where the reduction rate was maximum, the surface equiaxed grains were entirely transformed into fibrous microstructures. The deformed grains in the central layer were further flattened and elongated. The textures of the first-stage cold-rolled sheets were similar, including strong α, λ, and γ textures; however, the strong points and texture intensities were different. The reduction rate in process A was 40%, with the texture strength point located at {113}<230> (with an intensity of 4.26), and the deviation from {113}<361> was only 10°. The texture strength point changed with an increase in the reduction rate. In process B, the reduction rate was 50%, and the texture strength point was located at {113}<110> with an intensity of 4.79. When increased to 60% in process C, the texture strength point was located at {223}<110> and the intensity was 5.00. After annealing at 980 °C for 4 min, all deformed microstructures were completely recrystallized. With an increase in the reduction rate from 40% to 60%, the average grain size of the intermediate annealed sheets decreased gradually to approximately 85.6, 79.8 and 75.3 μm, respectively. When the reduction rate rose, the dislocation density in the cold-rolled sheet increased, which improved the driving force for recrystallization [22], and thus increased the nucleation ratio and decreased grain size of intermediate sheets. In process A, the texture was dominated by a strong γ texture and a small amount of α* texture, with a strong point near {111}<352> and an intensity of 4.08. In process B, the texture type changed slightly, whereas the strong point shifted to {332}<113>. In process C, the γ texture was weakened, the overall texture became dispersed, and the strong point was still located near {332}<113>.

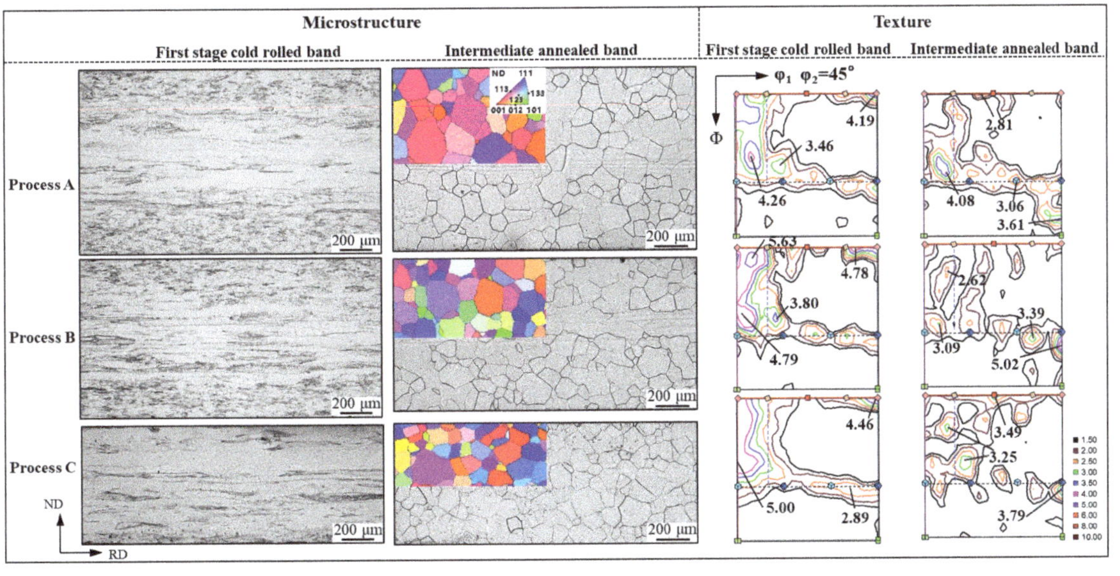

Figure 3. Microstructures and textures of first-stage rolled and intermediate annealed bands.

Figure 4 shows the microstructures and textures of the final cold-rolled sheets obtained by different processes. After single-stage cold rolling at a reduction ratio of 82.5%, the grains were severely compressed along the rolling direction, forming a fibrous microstructure.

Compared with the microstructure formed during single cold rolling, the microstructure of the double cold rolling process exhibited a significant change. Two kinds of deformed grains can be observed in the double cold-rolled samples, which are usually divided into dark regions (high energy storage and dislocation density) and bright regions (low energy storage and dislocation density) [23]. With a decrease in the secondary rolling reduction, the number of shear bands in the dark zone gradually decreased. When the secondary reduction ratio was reduced to 60%, the extent of cold deformation was the lowest, and most of the grains could adapt to the plane deformation process [24]; thus, the high fault density and high energy storage areas were reduced. The texture of single cold rolling was mainly composed of strong α- and γ-fiber textures, with the strong point located at the {223}<110> component, with the intensity of 8.62. The λ fiber texture was discontinuous and mainly concentrated in {001}<110>, and the orientation intensity was 7.27. The secondary cold-rolled sheets exhibited a heavy α-fiber texture and relatively slight γ-fiber and λ fiber textures. As for process A, the strong point was near the {223}<110> component with an intensity of 8.79. With the decrease in the double cold rolling reduction rate (processes B and C), the α-fiber texture gradually weakened and transformed into an α* texture, whereas the γ texture first became weaker and then stronger with the peak point converted from the {111}<110> to {111}<112> component. In terms of the λ texture, the strong points were all located at the {001}<120> component, and the orientation intensities first increased and then decreased, which were 5.57, 6.72, and 5.15, respectively.

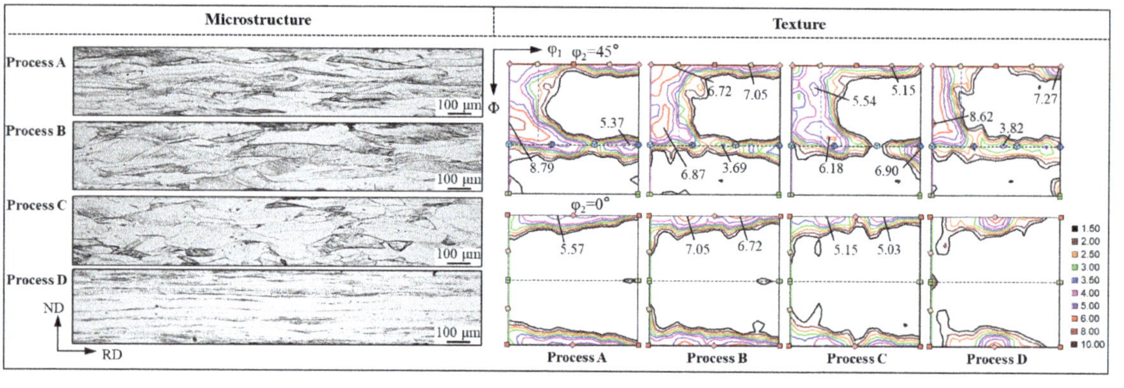

Figure 4. Microstructures and textures of the final cold-rolled bands.

3.3. Microstructures and Textures of Final Annealed Sheets

Figure 5 shows the microstructures and textures of the final annealed sheets. After final annealing of the single-stage cold-rolled sheet in process D at 750 °C, the rolled microstructure was fully recrystallized with grain size of approximately 28.2 μm. The single-stage cold-rolled sheet formed a strong γ and α* fiber texture, with the strong point located in the {111}<112> component (intensity 9.61). In the double cold rolling processes, all samples achieved complete recrystallization after final annealing with average grain sizes of ~32.2, ~38.7 and ~42.9 μm in processes A, B, and C, respectively. In contrast to single-stage rolling, the double cold rolling processes significantly reduced the adverse γ texture after final annealing. Process A showed a relatively strong γ fiber, α* fiber, and strong η fiber texture with a peak near the {114}<481> component. In process B, the γ fiber texture disappeared, while the η texture was significantly enhanced with an intensity of 6.25. In process C, the adverse γ-fiber texture appeared again, whereas the η texture was weakened.

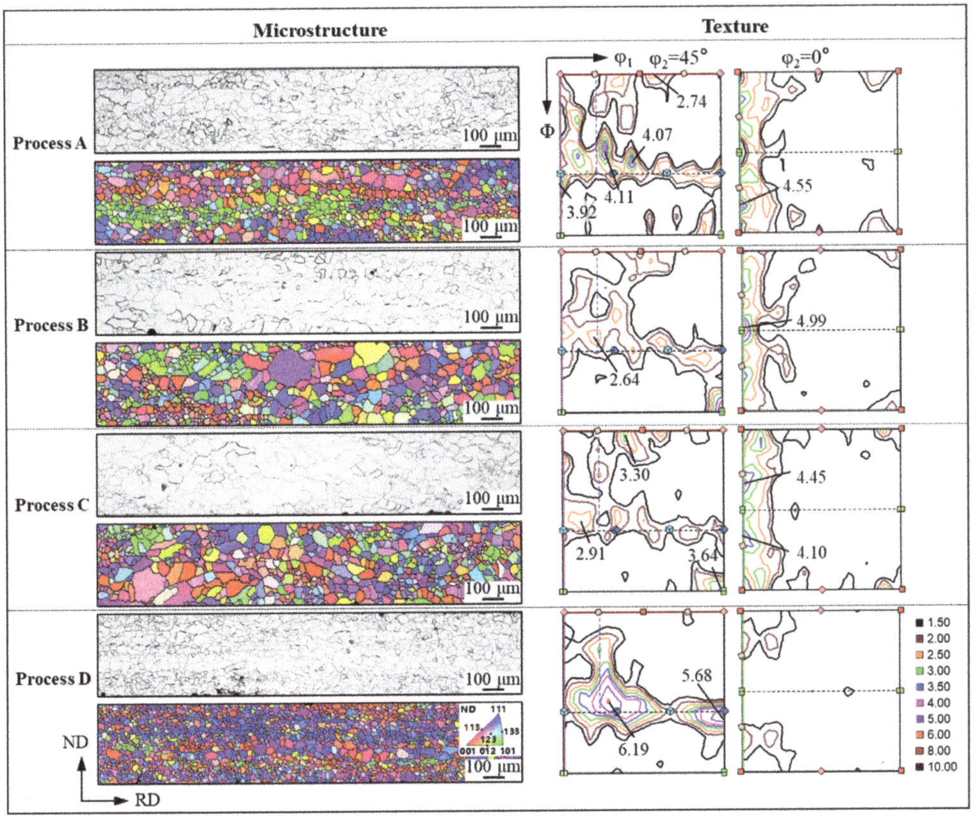

Figure 5. Microstructures and textures of the final annealed bands.

3.4. Magnetic and Mechanical Properties

Table 2 lists the magnetic and mechanical characteristics of Fe-3.3 wt% Si steel under different processes. The magnetic induction B_{50}, iron loss $P_{15/50}$, and $P_{10/400}$ of samples in process D were 1.660 T, 3.85 W/kg and 28.27 W/kg, respectively. In contrast, the magnetic properties of the bands prepared by the double cold rolling processes were apparently enhanced. From process A to C, B_{50} first went up and then went down. The trend of $P_{15/50}$ was opposite to that of B_{50}, first decreasing and then increasing. $P_{10/400}$ gradually decreased. When the reduction rates were 50% and 65% in process B, a good match between magnetic properties and mechanical properties was obtained (B_{50} = 1.683 T, $P_{15/50}$ = 2.96 W/kg, $P_{10/400}$ = 21.16 W/kg). Compared with single cold rolling, B_{50} increased by 0.023 T, $P_{15/50}$, $P_{10/400}$ and $R_{p0.2}$ decreased by 0.89 W/kg, 7.11 W/kg and 36 MPa, respectively.

Table 2. Magnetic and mechanical properties of Fe-3.3 wt% Si steel under different processes.

Samples	$P_{15/50}$ (W/Kg)	$P_{10/400}$ (W/Kg)	B_{50} (T)	$R_{p0.2}$ (MPa)	R_m (MPa)	E (%)
A-15m	3.26	21.46	1.671	508	612	19.4
B-15m	2.96	21.16	1.683	505	613	18.2
C-15m	3.07	21.15	1.679	496	594	20.1
D-15m	3.85	28.27	1.660	541	634	14.7

Table 3 shows the magnetic and mechanical properties of Fe-3.3 wt% Si steel annealed for various times under process B. When the annealing time was 200 s, B_{50}, $P_{15/50}$, and $P_{10/400}$ of the samples were 1.664 T, 4.40 W/kg and 29.42 W/kg, respectively. With the annealing time increasing to 5 min, B_{50} increased to 1.698 T, $P_{15/50}$ and $P_{10/400}$ decreased to 3.18 W/kg and 21.84 W/kg. When the annealing time was 10 min, the B_{50} of the sample decreased to 1.685 T and $P_{10/400}$ slightly increased to 21.97 W/kg. When the annealing time increased to 15 min, B_{50} slightly decreased, $P_{15/50}$ and $P_{10/400}$ decreased to 2.96 W /kg and 21.16 W/kg, respectively. With an increase in annealing time from 200 s to 15 min, the yield strength continuously decreased from 621 MPa to 505 MPa. Evidently, the magnetic and mechanical properties were well matched at 5 min, with B_{50} of 1.698 T, $P_{10/400}$ of 21.84 W/kg and $R_{P0.2}$ of 578 MPa. B_{50} increased by 0.038 T, $P_{10/400}$ reduced by 6.43 W/kg and the $R_{P0.2}$ increased by 37 MPa in comparison with the single-stage cold rolling (process D). Table 4 shows the property comparison between the Fe-3.3 wt% Si steel in this study and the industrial steel B35AHS550 (typical value) [25]. The mechanical properties of double cold-rolled Fe-3.3 wt% Si steel were at the same level as that of B35AHS550, while $P_{10/400}$ was reduced by 8.16 W/Kg and B_{50} increased by 0.038 T.

Table 3. Magnetic and mechanical properties of Fe-3.3 wt% Si steel under process B.

Samples	$P_{15/50}$ (W/Kg)	$P_{10/400}$ (W/Kg)	B_{50} (T)	$R_{p0.2}$ (MPa)	R_m (MPa)	E (%)
B-200s	4.40	29.42	1.664	621	671	10.6
B-250s	3.84	24.78	1.690	605	655	11.9
B-5m	3.28	21.84	1.698	578	648	13.4
B-10m	3.29	21.97	1.685	523	619	15.9
B-15m	2.96	21.16	1.683	505	613	18.2

Table 4. Comparison of properties of different silicon steels.

Samples	$P_{10/400}$ (W/Kg)	B_{50} (T)	$R_{p0.2}$ (MPa)
Fe-3.3 wt% Si steel in this work	21.84	1.698	578
Commercial B35AHS550	30.00	1.660	575

3.5. Relationship between Microstructure, Texture Evolution, and Properties

The texture of non-oriented silicon steel significantly influences the magnetic properties of the annealed sheets. This effect arises from the fact that the <001> direction of the BCC crystal structure exhibits the lowest magnetic crystal anisotropy energy, making it the preferred direction for magnetization. In contrast, the <111> direction represents a challenging magnetization direction. It is effective in improving the magnetic induction B_{50} of electrical steel by strengthening the λ and η textures (containing the most easily magnetized <100> direction), while minimizing the γ texture (containing the most difficult magnetized <111> direction) [4,26]. To investigate the correlation between various textures and magnetic characteristics, the texture factor (*TF*) [27,28], which is generally evaluated by the fraction of special textures, has been proposed. The η texture is considered in the texture factor, which is suitable for a strong η texture in this study. The texture factor can be expressed by the following formula:

$$TF = \frac{V_{<100>//ND} + V_{<100>//RD} - V_{\{100\}<001>}}{V_{<111>//ND}} \tag{1}$$

where V is the volume fraction of different textures, %. Since both λ and η fibers contain the Cube component, the volume fraction of the Cube texture needs to be subtracted here.

Figure 6 shows the texture volume fractions and texture factors of the final annealed sheets under different processes. The λ and η textures are beneficial to the magnetic susceptibility of B_{50}, whereas a stronger γ texture will worsen the magnetic induction B_{50} of silicon steel [29,30]. Obviously, the higher the TF value of the samples, the more favorable the magnetic properties [31]. Compared to the single-stage rolling process, the final annealed sheets prepared by double cold rolling contained lower γ and higher η and λ textures, leading to a higher texture factor and higher B_{50}. For the double cold rolling process A, B, and C, the volume fractions of the λ and η textures first increased and then decreased with a decline in the reduction ratio of double cold rolling. The volume fraction of the Cube texture was relatively low and its change was not obvious. The volume fraction of the γ texture gradually decreased. The above changes caused the texture factor to first increase from 1.32 to 2.13 and then decrease to 1.96, and all of the texture factors were higher than those of the single stage (0.48); therefore, B_{50} first went up from 1.671 T to 1.683 T and then went down to 1.679 T with the decrease in the reduction rate in the two-stage cold rolling processes, and all of the B_{50} were higher than that of the single stage (1.66 T). Iron loss is mainly influenced by grain size and texture. As the grain size increased, the density of grain boundaries diminished, which reduced the grain boundary pinning effect during magnetization, subsequently resulting in lower iron loss [32]. In addition, when the favorable textures in the annealed sheet increased, the energy required for magnetization was reduced, leading to the reduction in iron loss [33].The average grain sizes of process A, B, C, D after annealing for 15 min were ~32.2, ~38.7, ~42.9 and 28.2 μm, the texture factors were 1.32, 2.13, 1.96, and 0.48, and the corresponding $P_{10/400}$ were 21.46, 21.16, 21.15, and 28.27 W/kg, respectively. The grain sizes of process B and process C were similar and relatively large, and the texture factors were similar and relatively high; therefore, the iron loss $P_{10/400}$ of process B and process C were similar and the lowest.

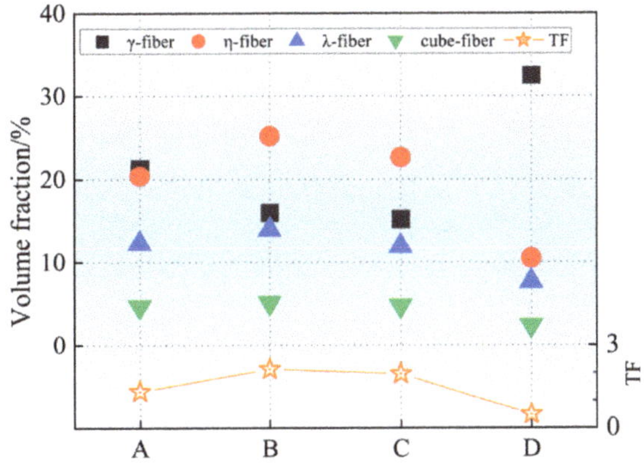

Figure 6. The volume fraction of specific textures and texture factors.

Figure 7 illustrates the microstructure and texture evolution in the final annealed sheet with different annealing times in process B. Figure 8 presents the evolution of the macro-texture in the final sheet. In the case of 200 s, the microstructure predominantly consisted of deformation bands. These deformation bands were predominantly composed of the γ- and α-fiber textures. Fine recrystallized grains were observed at the shear bands and boundaries of high-stored-energy γ-oriented grains, as well as at the boundaries of α-oriented grains. Recrystallized grains were largely absent in the α-oriented deformation bands, with nucleation occurring primarily at the grain boundaries. The newly formed grains did not grow significantly, and the overall texture primarily consisted of strong α- and γ-fiber textures, with the peak point located at {223}<110>. The recrystallization rate increased with increasing annealing time. At the annealing time of 250 s, recrystallized grains underwent some growth. Despite this growth, the overall microstructure changed slightly, with a significant presence of a deformation microstructure. The texture remained to be strong α- and γ-fiber components. The {223}<110> texture was significantly weakened, and the strong point was transferred to {118}<110>. A significant change in the microstructure was observed when the annealing time was extended to 5 min. The deformed grains were gradually consumed by the neighboring recrystallized grains, leading to a microstructure predominantly composed of recrystallized grains, with only a few incomplete deformed bands remaining. In addition, a strong component of the η fiber appeared in the sample, with the peak point located at the Goss texture and an intensity of 6.43. The γ fiber became dispersed and almost disappeared. During annealing from 10 min to 15 min, minimal changes in the microstructure and texture were noted. The microstructure gradually became homogeneous, while the texture continued to exhibit strong η-fiber characteristics, along with measurable intensities of the γ- and λ-fiber textures.

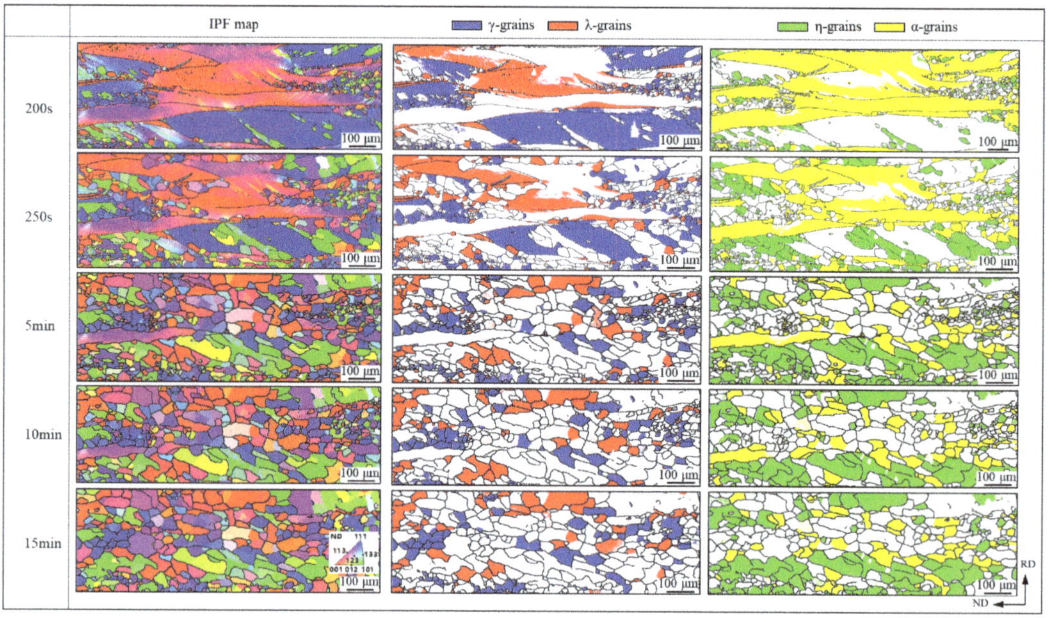

Figure 7. Microstructures and texture evolution of final sheet with different annealing times of process B.

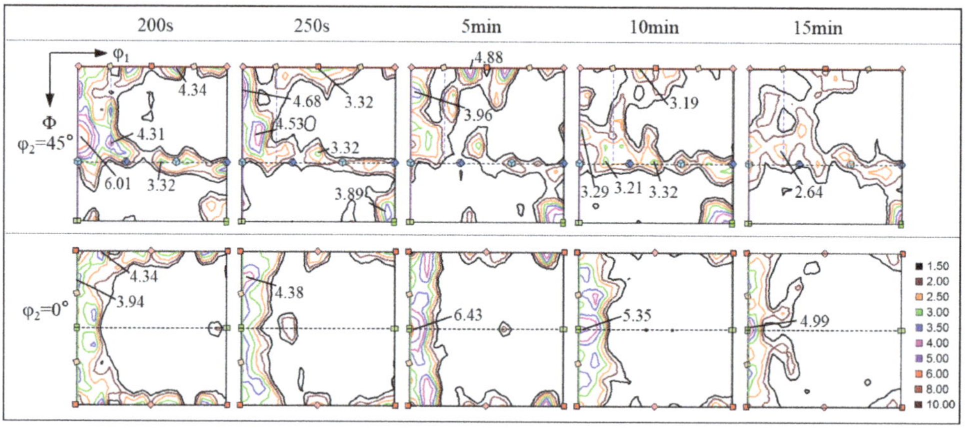

Figure 8. Macro-texture evolution of final sheet with different annealing times of process B.

Figure 9 shows the nucleation growth process and area fraction change in grains with specific orientations. Many studies [34,35] have shown that Goss grains mainly nucleate and grow in the shear zone of γ-oriented deformed grains. At positions 1 and 2, a Goss-oriented grain nucleated at the shear band of the {111}<112> deformed structure, whereas at position 3, the Goss-oriented recrystallized grain nucleated at the boundary of the {111}<112> deformed grain and remained during the subsequent annealing process. During the annealing process of 200 s~5 min, the Goss-oriented grains nucleated and grew, and the area fraction increased rapidly from 2.92% to 11.5%. However, after 10 min, some Goss grains were invaded by nearby grains (e.g., grains at position 7). The area fraction decreased to 11.0% and 10.7% at 10 min and 15 min, respectively. At position 4, the Cube-oriented recrystallized grains nucleated and grew at the shear band within the {111}<112> deformed structure, but the nucleation density was significantly lower than that of the Goss grains [36]. During the annealing process, the area fraction of Cube grains increased from 1.59% at 200 s, but was always low, with a maximum of 6.08%. At position 5, a {001}<120>-oriented grain nucleated at the boundary of the {111}<112> deformed grain and gradually grew into the interior of the deformed grain [12]. The area fraction of {001}<120> showed a decreasing trend from 7.66% to 4.18%. It is generally believed that γ-oriented grains usually nucleate and grow at the inner or grain boundaries of γ-deformed grains, resulting in the annealing texture of γ-fibers [37]. Position 6 shows that a {111}<110> grain nucleated at the boundary of the {111}<112> deformed structure. Position 7 shows that a {111}<112> recrystallized grain nucleated at the boundary of the {111}<110> deformed structure. During the annealing process, the area fraction of the {111}<112> oriented grains decreased from 19.9% to 4.92%, and then gradually increased to 5.76%. The area fraction of the {111}<110>-oriented grains showed a similar trend, decreasing rapidly from 8.28% to 2.61% and then gradually increasing to 5.6%.

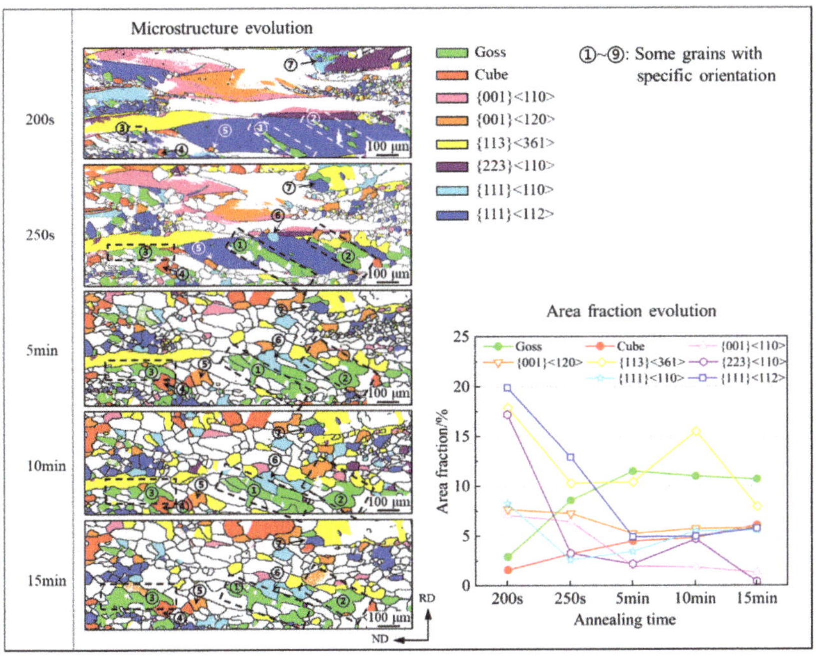

Figure 9. Microstructure and area fraction evolution of grains with specific orientations.

Figure 10 shows the volume fraction of specific textures and the texture factor in the final sheet for process B with different annealing times. Throughout the annealing process, the volume fraction of the Cube texture exhibited a slight increase. During annealing times of 200 s and 250 s, a significant number of deformed grains with γ orientation were observed. At an annealing time of 5 min, the deformed grains were progressively consumed by recrystallized grains with different orientations, leading to a rapid decrease in the fraction of the γ texture. The volume fractions of λ-fiber and η-fiber texture increased with the η-fiber texture component reaching its maximum, thereby achieving the highest texture factor. As γ-oriented grains nucleated and grew, the volume fraction of the γ-fiber texture increased again, while the dominance of λ and η textures diminished, causing a decrease in the volume fraction and a gradual reduction in the texture factor. Consequently, with increasing annealing time, the texture factor initially rose from 0.67 to 2.96 before subsequently declining to 2.13, and B_{50} exhibited a similar trend, increasing initially from 1.664 T to 1.698 T and then decreasing to 1.683 T. The texture factor was significantly increased compared with that of the single-stage rolling process (0.48); thus, B_{50} was improved up to 0.038 T.

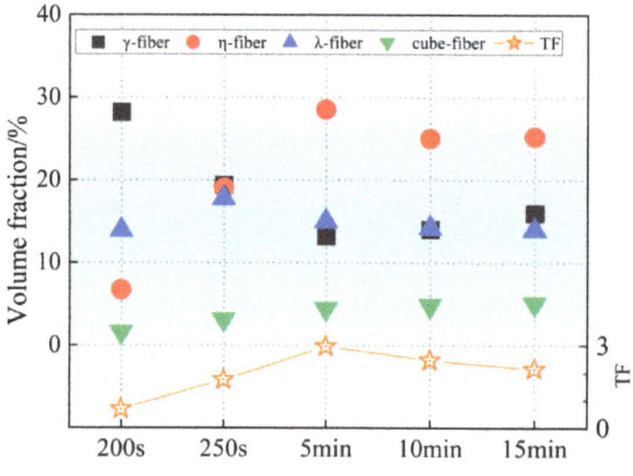

Figure 10. The volume fraction of specific textures and texture factors of final annealed sheet with different annealing times of process B.

As shown in Figure 7, short annealing time resulted in incomplete recrystallization of the sample. With increasing annealing time, the recrystallization rates gradually increased, which were 25.2%, 46.7%, 79.5%, 91.0%, and 97.6%, respectively. The incomplete recrystallized microstructures indicate that the sample contains many dislocations, and the distribution of dislocation density could be qualitatively represented by Geometrically Necessary Dislocations (GND). Figure 11 shows the GND maps of the final annealed sheet with different annealing times for process B. As the annealing time increased, the GND consistently decreased, indicating that dislocation density decreased gradually. Except for a slight increase at 10 min, the iron loss decreased with the increase in annealing time, which was mainly due to the increasing recrystallization, decrease in dislocation density, and possible grain growth. For similar reasons, with an increase in annealing time, the dislocation density decreased, resulting in a decrease in the effect of dislocation strengthening and a gradual decrease in yield strength. Compared to the single-stage process (~28.2 μm), although recrystallization was not complete, the average grain size of the sample was ~34.5 μm in the double-stage cold rolling process and annealing for 5 min, which finally reduced the iron loss $P_{10/400}$ by 6.43 W/kg. At the same time, a moderate dislocation density enhanced the yield strength by 37 MPa through dislocation strengthening.

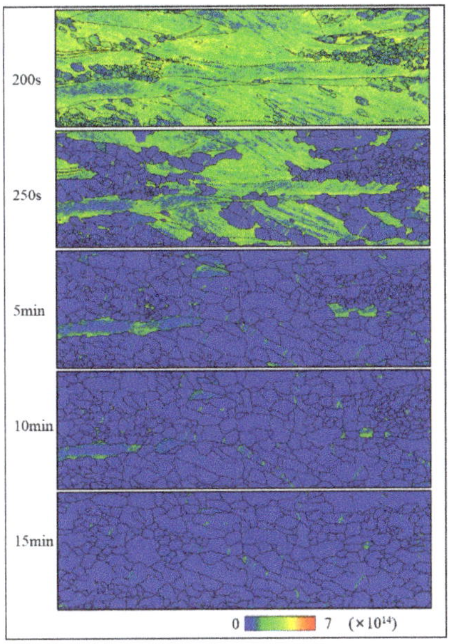

Figure 11. GND density of final annealed samples with different annealing times of process B.

4. Conclusions

Fe-3.3 wt% Si steel was prepared by different cold rolling processes and various annealing times. The main results are summarized as follows:

(1) The final annealed sheet produced through the single-stage cold rolling process is mainly composed of strong γ and α^* textures. In contrast, the double cold rolling processes significantly reduced the adverse γ texture, and a stronger η texture appears in the final annealed sheets, leading to a relatively high texture factor. Compared with single-stage cold rolling, the double cold rolling process can improve the magnetic induction and iron loss.

(2) Under the double cold rolling process, with the increase in annealing time, the magnetic induction B_{50} first increases from 1.664 T to 1.698 T and then decreases to 1.683 T, and the yield strength continues to reduce from 621 MPa to 505 MPa. With a shorter annealing time, incomplete recrystallization can improve the yield strength of the samples, owing to stronger dislocation strengthening.

(3) An optimized combination of magnetic properties (B_{50} of 1.698 T, $P_{10/400}$ of 21.84 W/kg) and mechanical properties ($R_{p0.2}$ of 578 MPa) is obtained via a moderate rolling reduction (50%, 65%) and an appropriate annealing time (5 min). Compared with the single-stage cold rolling process, B_{50} increases by 0.038 T, $P_{10/400}$ decreases by 6.43 W/kg and $R_{P0.2}$ increases by 37 MPa.

Author Contributions: Conceptualization, Y.G.; investigation, Y.G.; resources, Z.G.; software, H.C. and B.Y.; data collection, H.C. and B.Y.; data analysis, H.C. and B.Y.; writing—original draft preparation, Y.G.; writing—review and editing, Y.X., L.Z. and Z.G.; visualization, H.C. and B.Y.; funding acquisition, Y.X. and L.Z. All authors have read and agreed to the published version of the manuscript.

Funding: This research was supported by National Natural Science Foundation of China (Nos. 51974085 and 51674080), National Key R&D Program of China (2017YFB0304105, 2017YFB0304400), the Key R&D Program of Shandong Province (2019TSLH0103) and the Fundamental Research Funds for the Central Universities (N2425035).

Data Availability Statement: The data required to reproduce these findings cannot be shared at this time due to technical or time limitations.

Conflicts of Interest: Author Zhenyu Gao was employed by the company Silicon Steel Division of Angang Steel Co., Ltd. The remaining authors declare that the research was conducted in the absence of any commercial or financial relationships that could be construed as a potential conflict of interest.

References

1. Stornelli, G.; Faba, A.; Di Schino, A.; Folgarait, P.; Ridolfi, M.R.; Cardelli, E.; Montanari, R. Properties of Additively Manufactured Electric Steel Powder Cores with Increased Si Content. *Materials* **2021**, *14*, 1489. [CrossRef] [PubMed]
2. Leuning, N.; Jaeger, M.; Schauerte, B.; Stöcker, A.; Kawalla, R.; Wei, X.; Hirt, G.; Heller, M.; Korte-Kerzel, S.; Böhm, L. Material Design for Low-Loss Non-Oriented Electrical Steel for Energy Efficient Drives. *Materials* **2021**, *14*, 6588. [CrossRef] [PubMed]
3. Dou, Y.; Li, Y.; Zhang, C.; Yue, S.; Zhu, J. Effects of Uniaxial Stress Along Different Directions on Alternating Magnetic Properties of Silicon Steel Sheets. *IEEE Trans. Magn.* **2020**, *56*, 7512004. [CrossRef]
4. Du, Y.; O'Malley, R.; Buchely, M.F. Review of Magnetic Properties and Texture Evolution in Non-Oriented Electrical Steels. *Appl. Sci.* **2023**, *13*, 6097. [CrossRef]
5. Sahoo, G.; Singh, C.D.; Deepa, M.; Dhua, S.K.; Saxena, A. Recrystallization behaviour and texture of non-oriented electrical steels. *Mater. Sci. Eng. A* **2018**, *734*, 229–243. [CrossRef]
6. Ou, J.; Liu, Y.; Breining, P.; Gietzelt, T.; Wunsch, T.; Doppelbauer, M. Experimental Characterization and Feasibility Study on High Mechanical Strength Electrical Steels for High-Speed Motors Application. *IEEE Trans. Ind. Appl.* **2021**, *57*, 284–293. [CrossRef]
7. Tanaka, I.; Yashiki, H.; Iwamoto, S.; Takamaru, H.; Nakayama, T. Development of high strength electrical steel SXRC of resource-saving design. *Mater. Jpn.* **2011**, *49*, 29–31. [CrossRef]
8. Zhong, B.; Cheng, Z.; Volkova, O.; Zhang, X.; Liu, J. Effect of microstructure modification on magnetic and mechanical properties of high-grade non-oriented silicon steel during annealing treatment. *J. Mater. Res. Technol.* **2023**, *27*, 7730–7739. [CrossRef]
9. Yu, L.; Luo, H. Effect of partial recrystallization annealing on magnetic properties and mechanical properties of non-oriented silicon steel. *Acta Metall. Sin.* **2019**, *56*, 291–300.
10. Lin, Y.; Wang, H.; Wei, H.; Zhang, W.; Wang, S.; Qiu, S.; Lu, H.; Wang, Y. Thick-Gauge Ultra-High-Strength High-Silicon Non-Oriented Silicon Steel with Balanced Mechanical and Magnetic Properties Controlled by Partial Recrystallization Annealing. *JOM* **2022**, *74*, 3788–3798. [CrossRef]
11. Zhang, B.; Liang, Y.; Wen, S.; Wang, S.; Shi, X.; Ye, F.; Lin, J. High-strength low-iron-loss silicon steels fabricated by cold rolling. *J. Magn. Magn. Mater.* **2019**, *474*, 51–55. [CrossRef]
12. Takashima, M.; Komatsubara, M.; Morito, N. {001}<210> Texture Development by Two-Stage Cold Rolling Method in Non-oriented Electrical Steel. *ISIJ Int.* **1997**, *37*, 1263–1268.
13. An, L.-Z.; Wang, Y.-P.; Wang, G.-D.; Liu, H.-T. Comparative study on microstructure and texture evolution of low silicon non-oriented electrical steels along one-stage and two-stage cold rolling processes. *J. Magn. Magn. Mater.* **2023**, *567*, 170358. [CrossRef]
14. Qin, J.; Liu, D.-f.; Yue, Y.; Zhao, H.-j.; Lai, C.-b. Effect of normalization on texture evolution of 0.2-mm-thick thin-gauge non-oriented electrical steels with strong η-fiber textures. *J. Iron Steel Res. Int.* **2019**, *26*, 1219–1227. [CrossRef]
15. Yao, Y.; Sha, Y.; Liu, J.; Zhang, F.; Zuo, L. Texture and microstructure for magnetic properties of two-stage cold-rolled Fe-6.5 wt pct Si thin sheets. *Metall. Mater. Trans. A* **2016**, *47*, 5771–5776. [CrossRef]
16. Li, Z.-h.; Xie, S.-k.; Wang, G.-d.; Liu, H.-t. Ultrathin-gauge high silicon non-oriented electrical steel with high permeability and low core loss fabricated by optimized two-stage cold rolling method. *Mater. Charact.* **2022**, *183*, 111593. [CrossRef]
17. Lu, Y.; Zu, G.; Luo, L.; Wang, Y.; Gao, L.; Yuan, L.; Ran, X.; Zhang, X. Investigation of microstructure and properties of strip-cast 4.5 wt% Si non-oriented electrical steel by different rolling processes. *J. Magn. Magn. Mater.* **2020**, *497*, 165975. [CrossRef]
18. Mehdi, M.; He, Y.; Hilinski, E.J.; Kestens, L.A.I.; Edrisy, A. The evolution of cube ({001}<100>) texture in non-oriented electrical steel. *Acta Mater.* **2020**, *185*, 540–554.
19. Sanjari, M.; He, Y.; Hilinski, E.J.; Yue, S.; Kestens, L.A.I. Development of the {113}<uvw> texture during the annealing of a skew cold rolled non-oriented electrical steel. *Scr. Mater.* **2016**, *124*, 179–183.
20. Kim, T.-Y.; Kim, H.-K.; Jeong, Y.-K.; Ahn, Y.-K.; Shim, H.-S.; Kwon, D.; Hwang, N.-M. Ex-Situ Time Sequential Observation on Island and Peninsular Grains in Abnormally Growing Goss Grains in Fe–3%Si Steel. *Met. Mater. Int.* **2020**, *26*, 1200–1206. [CrossRef]
21. Fang, F.; Yu, C.; Wang, J.; Ding, Q.; Zhang, Y.; Wang, Y.; Kang, J.; Zhang, X.; Yuan, G.; Wang, G. Abnormal grain growth with preferred orientation in non-oriented silicon steel: A quasi-in situ study. *J. Mater. Sci.* **2024**, *59*, 3150–3167. [CrossRef]
22. Wang, Y.-Q.; Kong, W.-W.; Yuan, C.; Chen, Y.-P.; Liu, X.; Liu, S. Effects of annealing on microstructures and properties of cold-rolled GH3536 sheet. *Mater. Charact.* **2021**, *180*, 111409. [CrossRef]
23. Li, Z.-H.; Xie, S.-K.; Wang, G.-D.; Liu, H.-T. Dependence of recrystallization behavior and magnetic properties on grain size prior to cold rolling in high silicon non-oriented electrical steel. *J. Alloys Compd.* **2021**, *888*, 161576. [CrossRef]
24. Huang, Y.; Prangnell, P. The effect of cryogenic temperature and change in deformation mode on the limiting grain size in a severely deformed dilute aluminium alloy. *Acta Mater.* **2008**, *56*, 1619–1632. [CrossRef]
25. Qiu, S.; Niu, Y.; Qiao, J.; Wang, H. A Review of Development and Performance Control of Non-Oriented Silicon Steel for Drive Motors. *J. Anhui Univ. Technol. Nat. Sci.* **2024**, *41*, 350–364.

26. Jiao, H.; Xu, Y.; Zhao, L.; Misra, R.; Tang, Y.; Liu, D.; Hu, Y.; Zhao, M.; Shen, M. Texture evolution in twin-roll strip cast non-oriented electrical steel with strong Cube and Goss texture. *Acta Mater.* **2020**, *199*, 311–325. [CrossRef]
27. Qiao, J.; Liu, C.; Guo, F.; Xiang, L.; Qiu, S.; Wang, H. Effect of recrystallization annealing temperature on texture and magnetic properties of 2.97% Si non-oriented silicon steel. *Metall. Res. Technol.* **2019**, *116*, 412. [CrossRef]
28. Ni, Z.; Cheng, Z.; Jing, W.; Wendler, M.; Volkova, O.; Zhang, X.; Liu, J. Effects of Cu Content on Mechanical and Magnetic Properties of High-Strength Nonoriented Silicon Steel. *Steel Res. Int.* **2024**, 2300865. [CrossRef]
29. Chen, W.; Cheng, Z.; Wen, Q.; Wendler, M.; Volkova, O.; Liu, J. High-Strength Nonoriented Electrical Steel with Excellent Magnetic Properties Accomplished by Cu–Ni Multialloying. *Steel Res. Int.* **2024**, 2400254. [CrossRef]
30. Li, Z.-H.; Wang, G.-D.; Liu, H.-T. Effects of processing routes on recrystallization texture development and magnetic properties of 0.10 mm ultrathin non-oriented electrical steel sheets for high-frequency application. *J. Alloys Compd.* **2023**, *935*, 167984. [CrossRef]
31. Ghosh, P.; Chromik, R.R.; Vaseghi, B.; Knight, A.M. Effect of crystallographic texture on the bulk magnetic properties of non-oriented electrical steels. *J. Magn. Magn. Mater.* **2014**, *365*, 14–22. [CrossRef]
32. Leuning, N.; Steentjes, S.; Hameyer, K. Effect of grain size and magnetic texture on iron-loss components in NO electrical steel at different frequencies. *J. Magn. Magn. Mater.* **2019**, *469*, 373–382. [CrossRef]
33. Ning, X.; Liang, Y.; Wang, Y.; Ye, F.; Lin, J. A comprehensive study on texture evolution and recrystallization mechanism of Fe-4.5 wt% Si with strong {114}<481>. *J. Mater. Res. Technol.* **2023**, *27*, 7506–7520.
34. Liu, J.; Sha, Y.; Hu, K.; Zhang, F.; Zuo, L. Formation of Cube and Goss Texture After Primary Recrystallization in Electrical Steels. *Metall. Mater. Trans. A* **2014**, *45*, 134–138. [CrossRef]
35. Zhang, Y.; Xia, Y.; Dun, H.; Wang, Y.; Fang, F.; Zhang, Y.; Zhang, J.; Chen, Q.; Zhai, K.; Misra, R.D.K. Evolution of the Shear Band in Cold-Rolling of Strip-Cast Fe-1.3% Si Non-Oriented Silicon Steel. *Materials* **2021**, *14*, 775. [CrossRef]
36. Park, J.-T.; Szpunar, J.A. Evolution of recrystallization texture in nonoriented electrical steels. *Acta Mater.* **2003**, *51*, 3037–3051. [CrossRef]
37. Hawezy, D.; Birosca, S. Disparity in recrystallization of α- & γ-fibers and its impact on Cube texture formation in non-oriented electrical steel. *Acta Mater.* **2021**, *216*, 117141.

MDPI AG
Grosspeteranlage 5
4052 Basel
Switzerland
Tel.: +41 61 683 77 34

Metals Editorial Office
E-mail: metals@mdpi.com
www.mdpi.com/journal/metals

www.ingramcontent.com/pod-product-compliance
Lightning Source LLC
Chambersburg PA
CBHW062033220926
51850CB00010B/2537
* 9 7 8 3 7 2 5 8 7 4 9 6 5 *